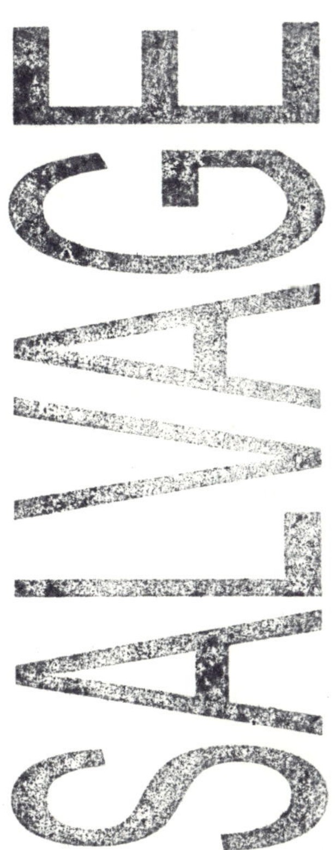
SALVAGE

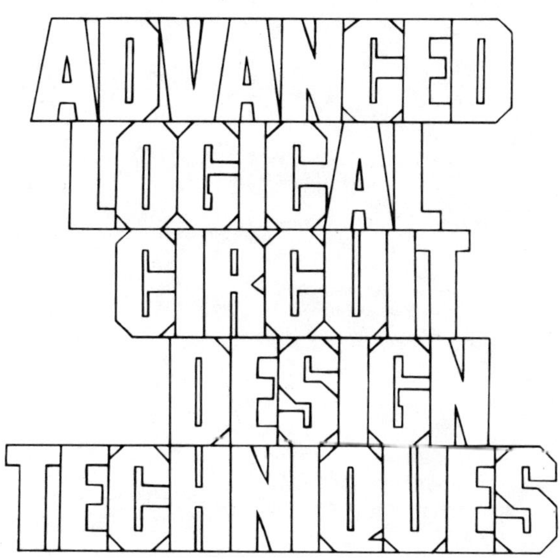

ADVANCED LOGICAL CIRCUIT DESIGN TECHNIQUES

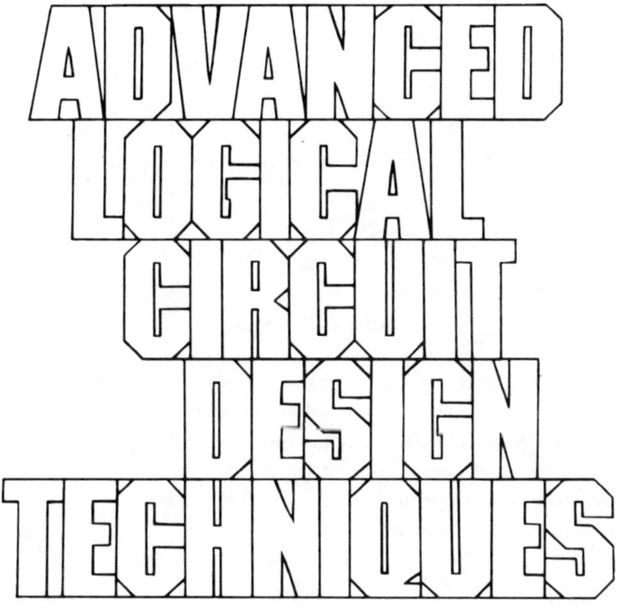

ADVANCED LOGICAL CIRCUIT DESIGN TECHNIQUES

Antonin Svoboda
University of California, Los Angeles

Donnamaie E. White
Advanced Micro Devices, Inc.

Garland STPM Press
New York & London

Library of Congress Cataloging in Publication Data

Svoboda, Antonín.
   Advanced logical circuit design techniques.

   Includes bibliographical references and index.
   1. Logic circuits.   2. Logic design.   3. Electronic circuit
design — Data processing.
I. White, Donnamaie E., 1942–   —joint author.
II. Title.
TK7868.L6S94   621.3815′3   78-31384
ISBN 0-8240-7014-3

Published by Garland STPM Press
545 Madison Avenue, New York, New York 10022

Printed in the United States of America

Contents

LIST OF FIGURES.. xi
LIST OF TABLES.. xvi
PREFACE...xvii

CHAPTER 1: COMPUTER-AIDED LOGIC DESIGN..............   1

1.1   Introduction.....................................   1
1.2   General Philosophy of Problem Specification
      and Solution....................................   2
      Example 1. Cause-effect and logical relation..   2
      Example 2. Marquand chart.......................   3
      Example 3. Logical distance.....................   6
      Example 4. Existence function -
        unconstrained system..........................   6
      Example 5. Existence function - constrained
        system...   7
      Example 6. Existence function - full adder....   7
      Example 7. Existence function - S-R NOR
        flip-flop.......................................   9
      Example 8. Equation formulation for BOOL...... 12

CHAPTER 2: GENERATION AND PROCESSING OF BOOLEAN
      FUNCTIONS....................................... 13

2.1   Introduction..................................... 13
2.2   Existence Function Generation................... 13
2.3   Truth Table Generation.......................... 15
2.4   Processing of Boolean Functions................. 15
2.5   Examples... 16
      Example 2.5.1 Developing the existence
        function of a full adder...................... 16
      Example 2.5.2 Full adder analysis based on
        its truth table............................... 19
      Example 2.5.3 Developing the Existence
        function of the S-R NOR flip-flop............ 23
      Example 2.5.4 Designing a combinational
        network from specification by truth table... 25
      Example 2.5.5 Minimizing the algebraic form
        of a function.................................. 28
      Example 2.5.6 Interchanging the indexing
        position of variables......................... 30

SECTION   PAGE

          Example 2.5.7 Finding the Boolean difference
          of a function............................ 32

CHAPTER 3: THE APL PROGRAM "SYSTEM"................. 35

CHAPTER 4: MINIMIZATION AND OPTIMIZATION........... 46

4.1   Introduction.................................. 46
4.2   Examples..................................... 47
      Example 4.2.1 Minimal sigma-pi form of a
      Boolean function............................. 47
      Example 4.2.2 Designing a two-level NOR
      network...................................... 50
      Example 4.2.3 Designing a full adder as a
      multiple output circuit...................... 51
      Example 4.2.4 Designing the full adder with
      a complemented output........................ 55
      Example 4.2.5 Minimization of CYCLIC and
      ABNORMAL Boolean functions................... 58
      Example 4.2.6 Minimization of CYCLIC and
      NORMAL Boolean functions..................... 60

CHAPTER 5: THE APL PROGRAM "OPTIMA"................. 63

CHAPTER 6: SEQUENTIAL CIRCUIT AIDES................. 76

6.1   Introduction.................................. 76
6.2   Designation of Functions..................... 76
6.3   Examples..................................... 77
      Example 6.3.1 Switching network transform
      (triangle to star)........................... 77
      Example 6.3.2 Finding steady states......... 80
      Example 6.3.3 Non-clocked J-K flip-flop
      without memory............................... 83
      Example 6.3.4 Non-clocked J-K flip-flop
      with memory.................................. 100

CHAPTER 7: THE APL PROGRAM "BOOL"................... 108

CHAPTER 8: MINIMIZATION TECHNIQUES.................. 113

8.1   Introduction.................................. 113
8.2   Single-Output Minimization................... 114
      8.2.1 Design constraints..................... 114
      8.2.2 Minimization by inspection............ 115

SECTION   PAGE

          8.2.3   Minimization by mapping and
                  observation.................................. 117
          8.2.4   Minimization by algebraic manipulation. 118
8.3   Svoboda's Weight Algorithm........................ 119
8.4   Logical Instruments: The Weight Deck........... 123
          8.4.1   Description of the cards............... 123
          8.4.2   Finding the weights for the
                  six-variable example...................... 123
8.5   Svoboda's Fundamental Product Procedure........ 125
          8.5.1   Introduction........................... 125
          8.5.2   The procedure for finding the
                  fundamental product and the effective mask.. 126

CHAPTER 9: INTRODUCTION TO TRIADIC GRAPHICAL
            CALCULUS.. 130

9.1   Introduction....................................... 130
9.2   Binary Space Notation............................. 130
9.3   Triadic Space Notation............................ 132
9.4   Correlations between Binary and Triadic Space.. 134
9.5   Triads... 134
9.6   Informational Content of Maps.................... 139
9.7   Boolean Notational Forms......................... 139
          9.7.1   Canonical sigma-pi-form............... 142
          9.7.2   Sigma-pi-form.......................... 143
          9.7.3   Maximal sigma-pi-form................. 144
          9.7.4   Complete sum........................... 145
9.8   The Triadic Map Algorithm for Finding the
      Complete Sum of y.................................. 146
9.9   The Petrick Function Solution for the
      Minimal Sigma-Pi Form of y....................... 148
9.10 Logical Instruments - Prime Implicant
      Generation.. 151
          9.10.1   Introduction......................... 151
          9.10.2   The minterm deck..................... 151
          9.10.3   The term deck........................ 154

CHAPTER 10: INTRODUCTION TO THE PARALLEL BOOLEAN
             PROCESSOR..................................... 157

10.1 Introduction....................................... 157
10.2 The Theorems....................................... 157
          10.2.1   Theorem 3.1.......................... 157
          10.2.2   Theorem 3.2.......................... 158
          10.2.3   Theorem 3.3.......................... 161
          10.2.4   Theorem 3.4.......................... 164
          10.2.5   Theorem 3.5.......................... 164
          10.2.6   Theorem 3.6.......................... 164
          10.2.7   Theorem 3.7.......................... 165

    10.2.8   Theorem 3.7A.......................... 165
10.3 The Original Design of the Parallel Boolean
    Processor.. 167
    10.3.1   Introduction........................... 167
    10.3.2   The original design.................... 168
    10.3.3   Improved parallelism in the
    processor implementation..................... 172
10.4 Applications...................................... 177
    10.4.1   Implicant listing..................... 177
    10.4.2   Implicant listing under improved
    parallelism.................................. 177
    10.4.3   Existence function.................... 181
    10.4.4   Larger systems....................... 181
        10.4.4.1   More terms...................... 181
        10.4.4.2   More variables................. 183
10.5 The Coverage Algorithm........................... 183
10.6 The Boolean Difference........................... 188

CHAPTER 11: DESIGNING WITH MSI-LSI.................... 202

11.1 Introduction..................................... 202
11.2 SSI Design....................................... 204
11.3 Gate versus Connection Minimization............. 205
11.4 MSI Design....................................... 210
11.5 LSI Design Techniques............................ 218
    11.5.1   Programmable multiplexers............. 219
    11.5.2   Programmed logic arrays............... 219
    11.5.3   Programmable array logic............. 223
    11.5.4   Design example........................ 225
    11.5.5   Read-only memories.................... 230
    11.5.6   Sequential design example............. 233

CHAPTER 12: FAULT DETECTION TECHNIQUES............... 241

12.1 Faults... 241
    12.1.1   Fault definition...................... 241
    12.1.2   Masking a fault....................... 241
    12.1.3   Fault types........................... 243
    12.1.4   Fault equivalencies................... 243
    12.1.5   The problem........................... 244
12.2 The Test Sequence................................ 246
    12.2.1   Deriving the Existence Function....... 246
        12.2.1.1   The equations................... 246
        12.2.1.2   Required labeling............... 247
        12.2.1.3   Equation reformulation........ 249
        12.2.1.4   Generating the Existence
        Function..................................... 250
    12.2.2   Deriving the Test Sequence............ 251
        12.2.2.1   Formation rules................. 251

12.2.2.2   Chain selection................ 251
12.2.2.3   Advantages of the Test
Sequence................................. 254
12.2.2.4 Possible extension to
sequential circuits.................... 258
12.3 Examples.. 258
12.3.1   Elementary gates..................... 258
12.3.2   Test sequence versus Boolean
Difference............................... 260
12.3.3   A diagnostic table................... 260
12.3.4   Equivalent circuits: Test Sequence
versus Kohavi's Maps..................... 265
12.3.5   Multiple faults..................... 269
12.4 Summary... 269

REFERENCES.. 275

APL Program INDEX.................................... 279

List of Figures

FIGURE   PAGE

CHAPTER 1
1.1   A Marquand Chart for a Five-Variable System...   5
1.2   Examples of Existence Functions...............   6
1.3   A Combinational Circuit.......................   7
1.4   Truth Table of the Full Adder.................   8
1.5   NOR Flip-Flop With Reset......................   9
1.6   State Transistions for Flip-Flop of Figure 1.5   10

CHAPTER 2
2.1   Existence Function Generation.................   14
2.2   Truth Table Generation........................   14
2.3   Program Module Sequence for Truth Table of
       Threshold Functions...........................   19
2.4   Program Module Sequence for the Existence
       Function of a Sequential Circuit..............   23
2.5   Program Module Sequence for Combinational
       Design via Truth Table........................   25

CHAPTER 3 (NONE)

CHAPTER 4
4.1   Program Module Sequence for Minimization and
       Optimization..................................   46

CHAPTER 5 (NONE)

CHAPTER 6
6.1   Switching Network of Example 6.3.1............   77
6.2   Marquand Map of Example 6.3.1 - Solution No.1.   79
6.3   Sequential Circuit of Example 6.3.2...........   80
6.4   Cause-Effect Chain............................   83
6.5   Transition Diagram............................   84
6.6   Dynamic Schematic: J-K Flip-Flop - Trailing
       Edge Controlled - No Input Inverters..........   93
6.7   Dynamic Schematic: Rods Placed to Represent
       the Steady State $\underline{A}\ \underline{B}\ C\ D\ E\ F = 1$..............   95
6.8   Dynamic Schematic: $\overline{J}$-K Flip-Flop - Trailing
       Edge Controlled - Input Inverter Hazards......   98

6.9   Dynamic Schematic: J-K Flip-Flop – Trailing
       Edge Controlled – Hazards Eliminated.........   99
6.10   Sequential Circuit with Memory Elements (S-R
       Flip-Flops) in the Feedback Loop.............. 100
6.11   State Diagram for Constraints Generation...... 103
6.12   J-K Flip-Flop With Trailing Edge Control...... 107
6.13   Final Design Version of J-K Flip-Flop......... 107

CHAPTER 7 (NONE)

CHAPTER 8
8.1   Combinational, Single-Output Circuit.......... 116
8.2   Reducing a Simple Function.................... 117
8.3   Algebraic and Map Minimization of a
       Five-Variable Function........................ 118
8.4   The Weight Algorithm for a Five-Variable
       Function...................................... 121
8.5   The Weight Algorithm for a Six-Variable
       Function...................................... 122
8.6   Weight Deck Card A0........................... 124
8.7   Fundamental Product Solution.................. 127

CHAPTER 9
9.1   Marquand Map for the Function $m_a = X_3\overline{X_2}X_1\overline{X_0}$
       ($m_a = m_{10}$).............................. 132
9.2   Triadic Map with Minterm Space Labeled with
       Point Identifiers............................. 133
9.3   Maps and Their Identifiers.................... 135
9.4   Structures in Binary and Triadic Space........ 136
9.5   Spatial Relationships......................... 137
9.6   Example Triads................................ 138
9.7   Detail of Term Relationships for a
       Four-Variable Map............................. 141
9.8   Triadic Map Symbol Intersection Chart......... 142
9.9   Canonical Sigma-Pi-Form of an Example Function 143
9.10   Map of a Sigma-Pi-Form of y................... 144
9.11   Map of the Maximal Sigma-Pi-Form of y......... 145
9.12   Map of the Complete Sum of y.................. 146
9.13   Finding the Complete Sum...................... 147
9.14   Finding the Prime Implicants via the Triadic
       Map... 149
9.15   Prime Implicant Table......................... 150
9.16   Map of Minterm Card 4......................... 152
9.17   Example Function.............................. 153
9.18   The Maximal Sigma-Pi-Form of y................ 153
9.19   Map of Triadic Card 10........................ 154
9.20   The Complete Sum.............................. 155

CHAPTER 10
10.1   Theorem 3.1................................... 159
10.2   Theorem 3.1 in Complement Space............... 160

FIGURE   PAGE

10.3   Theorem 3.2................................... 162
10.4   Theorem 3.3................................... 163
10.5   Theorem 3.7................................... 166
10.6   Block Diagram - Parallel Boolean Processor.... 169
10.7   Processing Register Detail.................... 170
10.8   Original Design of the Two Lower Digits for
       the Processing Registers...................... 171
10.9   The Functions ϕ for a Clock Step of 3**2...... 173
10.10 Processing Register with 9 Parallel Outputs... 174
10.11 Map of o_{q*} Where i = 3...................... 175
10.12 The o_{q*} Functions for a 3^3 Clock Step........ 176
10.13 Implicant Listing - Example Clock Step....... 178
10.14 Detail of Implicant Listing.................. 179
10.15 Implicant Listing - Output and Register
       Contents...................................... 180
10.16 Generating the Existence Matrix.............. 182
10.17 Block Diagram - Parallel Boolean Processor.... 184
10.18 Map and Clock Step Detail of MULTIPLICITY..... 186
10.19 Clock Step Detail of COVERAGE................. 187
10.20 Boolean Difference by Computation and
       Mapping....................................... 189
10.21 An Alternative Map Approach................... 190
10.22 Solution with Links and the Existence
       Function Map.................................. 192
10.23 Parallel Boolean Processor Algorithm for
       Implementation of Map Approach................ 194
10.24 Parallel Boolean Processor Algorithms for a
       More Complex Example.......................... 196
10.25 The Solution with Links....................... 198
10.26 Another Example Difference from Sellers....... 200

CHAPTER 11
11.1   Minimal Pi-Sigma-Forms of y_1 and y_2.......... 206
11.2   Comparison of SSI Implementations............. 208
11.3   An EXOR Implementation of y_1................. 209
11.4   One of Four Multiplexer....................... 211
11.5   Implementation of y_1 and y_2 with Multiplexers. 212
11.6   The Four Variable Function y_{EBFF} from
       Muruga and Lai's Paper........................ 214
11.7   Multiplexer Implementation of a Multiple
       Output Problem................................ 215
11.8   Svoboda's Six-Variable Example Done with
       Multiplexers.................................. 216
11.9   Comparison of Implementations of the
       Six-variable Example.......................... 217
11.10 Designing with a Small PLA.................... 221
11.11 Implementation of the Six-Variable Problem
       with a PAL.................................... 224
11.12 Basham's Modified Hamming Code................ 226
11.13 Parity and Check Bit Generation............... 227
11.14 Syndrome Bit Generation with Multiplexer...... 228

FIGURE   PAGE

11.15 Syndrome Bit Decode............................ 228
11.16 Control Signal Generation...................... 229
11.17 Programmed PAL for Control Signal Generation.. 231
11.18 ROM Logic Block................................ 232
11.19 Multiple Output Problem Implemented in RCM.... 232
11.20 Svoboda's Six-Variable Problem done with a
       PROM...................................... 234
11.21 Traffic Light Controller....................... 235
11.22 Light Control Signal State Sequencing......... 237
11.23 Traffic Light Control Design with Am2910...... 238
11.24 Microprogram for the Controller............... 239

CHAPTER 12
12.1   Redundant Circuit.............................. 242
12.2   Sample Circuit for Comparisons................ 245
12.3   Examples of Labeling; Equations............... 248
12.4   Existence Function Generation................. 252
12.5   Link Formation................................ 253
12.6   The Test Sequence............................. 255
12.7   Diagnostic Continuity Diagram................. 257
12.8   Test Sequences for Elementary Gates........... 259
12.9   An Example from Marinos........................ 261
12.10 An Example from Bearnson and Carroll......... 263
12.11 Tabular Sequence Generation................... 264
12.12 An Example from Kohavi and Kohavi............. 266
12.13 Three Example Circuits from Kohavi and Kohavi. 267
12.14 Output from BOOLE for Figure 12.12............ 268
12.15 Output from BOOLE for Figure 12.13c........... 268
12.16 Multiple Fault Testing Example from Yau and
       Tang...................................... 270
12.17 Diagnostic Continuity Diagram for Figure 12.16 271
12.18 Fault Table Test Set Generation for Figure
       12.16..................................... 272

List of Tables

TABLE   PAGE

2.1   Routines for Processing Boolean Functions..... 15
6.1   Constraint Formulations for Example 6.1.3..... 87
6.2   Constraint Formulations for Example 6.1.2..... 102
8.1   Design Constraints............................ 116
9.1   Binary Map Symbols............................ 140
9.2   Triadic Map Symbols.......................... 140
11.1   Example Products............................. 218
11.2   Table of Am2910 Instructions Used............ 240

Preface

This text is the compilation of courses developed
by Antonin Svoboda and presented by him while he was
Professor Emeritus of the Computer Science Department
of the University of California, Los Angeles, and of
courses developed by one of his graduate students,
Dommamaie E. White, and presented by her while she was
on the part-time faculty of the School of Engineering,
California State University, Los Angeles. The material
was first combined and presented by its authors at a
professional seminar held by the University Extension
in early 1977.

Over his lifetime, Svoboda has pursued many in-
terests. Those aspects of his work which are represented here are concerned with his Parallel Boolean
Processor and with his theorems and his unique approaches toward finding the minimal or optimal solutions to fundamental combinational and sequential circuit design problems. The Triadic notation which
appears heavily throughout many of his published
papers is clearly documented for the first time.

Dr. Svoboda built the world's first fault tolerant computer, SAPO, for the Academie of Science in
Prague. Designed in 1950, SAPO was operational in 1954.

Chapters 3, 5, and 7 of this text present the
complete listings of the APL circuit laboratory which
Svoboda created while he was at UCLA. A number of
examples are included in the surrounding chapters to
demonstrate the application of the various program
modules.

The remaining chapters contain detailed explanations and examples of various design problems from
minimization of single output combinational functions
through the mosaics of multiple output functions. An
explanation of the Parallel Boolean Processor and its
fundamental theorems is offered without any attempt
to duplicate completely the material included in the
referenced published papers. Rather, the intent is
to clarify the earlier papers which had been edited
to meet space restrictions at the time that they were
published. The new, unpublished COVERAGE algorithm is
described.

Applications of the Parallel Boolean Processor are implied, with several new applications presented in detail. These include the development of the Test Sequence for fault detection testing of combinational circuits.

The minimization techniques: 1) the weight algorithm; 2) fundamental product; 3) mosaics of functions and 4) coverage are representations of the application areas of Svoboda's theorems. In fact, the techniques described in chapters 8-11 are verbal descriptions of the APL program library presented in chapters 1-7. The APL package was written using triadic notation.

Svoboda's approaches are unique and elegant in their simplicity. The Marquand map, proposed in 1881 and overlooked by logic designers until recently, and the Triadic map developed by Svoboda are the tools which he uses to graphically explain his theorems and techniques. These are the "missing links" which anyone attempting to study the basic fundamentals of Boolean logic will find invaluable. They lend themselves readily to algorithmic manipulation via APL.

This book is intended as a reference text rather than as a text book per se although the material has been combined with appropriate exercises and lecture material and used by both of the authors in both undergraduate and graduate courses. Its primary function is to document the extensive APL circuit laboratory and Svoboda's techniques for the benefit of those who were not privileged enough to have attended his seminars.

Donnamaie E. White

Chapter 1
Computer-Aided Logic Design

1.1   INTRODUCTION

Hardware components of computers are physical models of logical reasoning. Procedures based on logical disciplines of mathematics are used to design these components. Examples of such procedures will be presented here in the form of APL programs intended to solve basic problems of computer logic design.

The three blocks of programs given here were used by the participants of the Short Course, Advanced Logical Circuit Design Techniques, presented by UCLA Extension (March, 1977). They are:

1.   SYSTEM -- permits the transformation of a problem specification into a set of Boolean functions defined by a truth table; derives the existence function of the system; and provides a tool for minimization and for logical relation analysis.

2.   OPTIMA -- permits the optimal design of a two-level multiple-output combinational circuit based on a rigorous mathematical principle.

3.   BOOL -- solves systems of Boolean equations of the general type; used for computer-aided design of sequential circuits.

CHAPTER 1

To use the programs effectively, it is necessary that one understand (1) the general philosophy of problem specification and solution; (2) the programming symbolism; and (3) the interpretation of the printout.

1.2 GENERAL PHILOSOPHY OF PROBLEM SPECIFICATION AND SOLUTION

The following unified point of view is recommended for solving problems in logic design:

1. SPECIFICATIONS should be presented in propositional calculus, Boolean algebra, or in the algebra of sets (classes);

2. EXECUTION of the solution procedure should be based on the algebra of sets (chart methods);

3. RESULTS are represented formally in Boolean algebra or by graphics.

The hardware design deals with physical phenomena related by a CAUSE-EFFECT relationship between STATES (events). The STATE of a SYSTEM can be described as a configuration of validities of propositions concerning measurable quantities (i.e., voltages, currents, etc.).

EXAMPLE 1. For voltage measurement of terminals X1, X2, ... , XJ, ... we use PROPOSITIONAL VARIABLES (x 1), (X 2), ... , (X J), ... attached to the propositions describing the outcome of voltage measurement:

$$(X\ J) \equiv (\text{Terminal XJ is HIGH}) \equiv (\text{Voltage at XJ is above 4.5 V})$$
$$(\underline{X}\ J) \equiv (\text{Terminal XJ is LOW}) \equiv (\text{Voltage at XJ is below 0.5 V})$$

Note: (1) The existence of two thresholds and their separation; (2) $(\underline{X}\ J)$ is a negation of (X J) so that [(X J) is TRUE] => [$(\underline{X}\ J)$ is FALSE] and vice versa;

(3) <u>underlining</u> of literals is used to express NEGATION (complementation).

The TRANSITION of a system from one state to another will be described by TWO SUBSEQUENT STATES. The first will be called the CAUSE of the state which follows it, which will be called its EFFECT.

LOGICAL TIME will be defined later and the definition will be derived from the CAUSE-EFFECT relationship previously mentioned.

A SUBSYSTEM is a subset of variables of a system that possesses certain properties. For instance, input variables of a combinational circuit have the property that they are mutually independent, and the output variables of the circuit have the property that each one is a Boolean function of the input variables (exclusively). In this case, we have two subsystems within a system. There may be more than two subsystems to consider when solving some problems of circuit design.

The LOGICAL RELATION between subsystems belonging to a system will be explained here for <u>two</u> subsystems by using Marquand charts of Boolean functions.

EXAMPLE 2. Subsystem with X-variables: (X J); J = 1,2,3 and the subsystem with Y-variables: (Y K); K = 1,2 form a SYSTEM. The validities of X-variables can take on eight different configurations; the validities of Y-variables can take on four configurations. When there is no logical relation between the subsystems, each of the 32 validity configurations of all five variables of the system is equally possible. When the system obeys postulated conditions (constraints), there will be a set of configurations (here from the set of 32) that will be ruled out (discarded). The configurations that survive the process of elimination define the EXISTENCE FUNCTION of the system as a whole.

CHAPTER 1

    To describe the Marquand chart suitable to explain
the concept of logical relation, the configurations of
validities of X- and Y-variables are identified (labeled)
by integers in the usual way.

    The eight possible configurations of validities of
X-variables will be identified by the integer IX, where
IX $\in$ {0,1,2,3,4,5,6,7} under the rule that IX, written
as a three-bit binary number $(x_3\ x_2\ x_1)_2$, belongs to the
configuration of validities: (X 3) = x_3, (X 2) = x_2,
(X 1) = x_1. For instance, IX = 3 = $(0\ 1\ 1)_2$ stands for
(X 3) = 0 (false) and (X 2) = (X 1) = 1 (true).

    Similarly, the four possible configurations of vali-
dities of Y-variables will be identified by the integer
IY, IY $\in$ {0,1,2,3}. When IY = 3 = $(1\ 1)_2$, both variables
(Y 1) = (Y 2) = 1 (true).

    The Marquand chart for the system of our example is
shown in Fig. 1.1. The horizontal scale of the chart
belongs to the subsystem X: (X J); J = 1,2,3; NX = 3.
The columns are labeled in IX from left to right, IX =
0,1,2,3,4,5,6,7; the number of columns (total number of
validity configurations) is designated by NNX, NNX = 8.
The vertical scale of the chart belongs to the subsystem
Y: (Y K); K = 1,2; NY = 2, NNY = 4 (number of rows).
The rows are labeled in IY = 0,1,2,3.

    Marquand conceived his chart in a binary way (in
agreement with the labeling practices of today). Written
as binary numbers, the identifiers IX, IY produce the
variables' validity configurations immediately. The
window of the chart (IY, IX) in the row IY and the column
IX corresponds to the validity configuration (Y2, Y1,
X3, X2, X1) of all five variables of this system. The
identifier of that configuration IS = $(y2\ y1\ x3\ x2\ x1)_2$ =
8×IY+IX. Each window in Fig. 1.1 is labeled with the
corresponding value of IS.

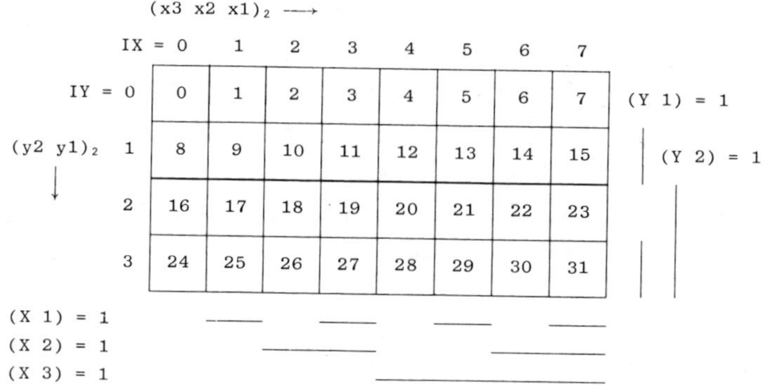

Figure 1.1. A Marquand chart
for a five-variable system.

The values of IS follow each other in a natural way.
This rule holds for a Marquand chart of any dimension and
any shape. The logical distance of a pair of windows on
the chart is the sum of disagreements in bits of their
binary identifers, IS. Two windows that are at the logical distance of one unit possess identifier IS values
that differ by 2^k (where k is an integer), thus implying
that two windows that are at the logical distance of one
unit must fall both in the same row or both in the same
column. Finally, the binary background of the chart leads
to the following simple rule: If a Marquand chart of any
size or of any shape is divided into vertical bands of
equal width 2^{k+1}, then any two windows within the same
band possessing the horizontal distance of 2^k (half of
band) have a logical distance of one unit. The same rule
holds for division into horizontal bands of equal width,
2^{k+1}. Any two windows in the <u>vertical</u> distance of 2^k
(both being in the same column, of course) that fall in
the same horizontal band have the logical distance of one
unit.

EXAMPLE 3. Four vertical bands in Fig. 1.1 have the width $2^1 = 2$ windows. For that reason, any two windows at the horizontal distance of 2^0 (1 window) falling in the same band have the logical distance of one unit; for instance, pairs of windows labeled in IS: (0,1), (12,13), (26,27). But pair (21,22), which has a horizontal distance of one window, is composed of elements that do not fall in the same band; their logical distance is not equal to one unit. Horizontal band division with band width 2 shows that (21,29) are windows at the logical distance of one unit, but that windows (10,18) are not. Vertical band division with band width 4 indicates that (17,19) are at the logical distance of one unit and that (19,21) are not.

Returning back to the LOGICAL RELATION between subsystems, four examples are offered.

EXAMPLE 4. Fig. 1.2a shows the existence function of a system whose subsystems X, Y are completely independent. The system X ∪ Y is without constraints. The chart is filled with "1"s to express that every possible validity configuration <u>exists</u>.

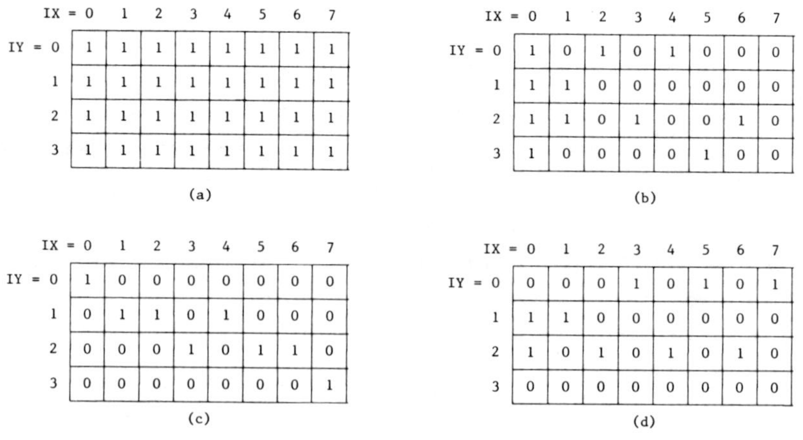

(a)

IX =	0	1	2	3	4	5	6	7
IY = 0	1	1	1	1	1	1	1	1
1	1	1	1	1	1	1	1	1
2	1	1	1	1	1	1	1	1
3	1	1	1	1	1	1	1	1

(b)

IX =	0	1	2	3	4	5	6	7
IY = 0	1	0	1	0	1	0	0	0
1	1	1	0	0	0	0	0	0
2	1	1	0	1	0	0	1	0
3	1	0	0	0	0	1	0	0

(c)

IX =	0	1	2	3	4	5	6	7
IY = 0	1	0	0	0	0	0	0	0
1	0	1	1	0	1	0	0	0
2	0	0	0	1	0	1	1	0
3	0	0	0	0	0	0	0	1

(d)

IX =	0	1	2	3	4	5	6	7
IY = 0	0	0	0	1	0	1	0	1
1	1	1	0	0	0	0	0	0
2	1	0	1	0	1	0	1	0
3	0	0	0	0	0	0	0	0

Figure 1.2. Examples of existence functions.

EXAMPLE 5.   Fig. 1.2b shows the existence function of a system subjected to some constraints.   In general, a given existence function can belong to many different sets of constraints.   We will mention the most obvious:

1.   For IX $\in$ {2,3,4,5,6}, the value of IY is uniquely determined.   In other words, IY is a function of IX within that domain.

2.   For IX = 1, it is IY = 1 XOR 2 (exclusive OR).   In another form,

$$(IX = 1) \rightarrow (Y2 \neq Y1).$$

3.   For IX = 0, it is IY = (any).   In other words,

$$(IX = 0) \rightarrow (\text{any one from all})$$
$$\equiv (\text{don't care which IY}).$$

4.   The input configuration belonging to IX = 7 is forbidden as the circuit has no steady state for X3 = X2 = X1.

EXAMPLE 6.   Fig. 1.2c shows the existence function of the FULL ADDER (Fig. 1.3).   It is a combinational circuit: A definite output signal configuration belongs to any input signal configuration.   In other words, the chart of the existence function must have exactly <u>one</u> non-zero in each column.   In another form, IY = f(IX). We say that the subsystem Y is a <u>function</u> of the subsys-

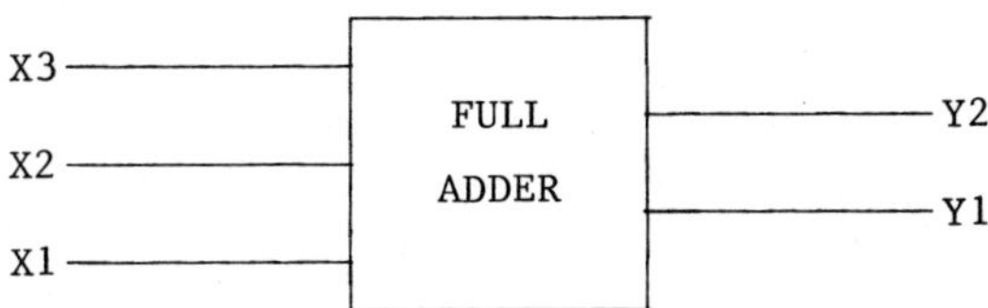

Figure 1.3.   A combinational circuit.

tem X. Symbollically, IX → IY. The constraint for the full adder, written in APL, is:

$$((Y\ 1)\ +\ 2\times(Y\ 2))\ =\ (X\ 3)\ +\ (X\ 2)\ +\ (X\ 1). \qquad (1.1)$$

The equation means that the sum $\sum_J$ (X J) (count of HIGHs at the input of the full adder) is equal to the binary number $(y2\ y1)_2$ (represented by HIGHs at the outputs).

It is important to point out that Fig. 1.2c is the chart of the <u>existence</u> <u>function</u> of the full adder and not the truth table of functions generated by the full adder. The relation between the existence function and the truth table is very simple:

1. The existence function can be <u>replaced</u> by a truth table <u>uniquely</u> if and only if each column of the (normalized) chart of the existence function contains exactly one non-zero.

2. The truth table function $((Y\ K)\ =\ 1)\to(Z\ K)$ can be deciphered from the existence function by reading IX values for which (Y K) = 1.

To get the truth table of the full adder from its existence function in Fig. 1.2c we start with (Y 1) = 1 to get (Z 1). Configurations with (Y 1) = 1 are all in the rows IY $\in$ {1,3}, and the existence function indicates that only four cases exist with IX $\in$ {1,2,4,7}, so that (Z 1) $\equiv$ (01101001). Similarly, (Y 2) = 1 is true only

$$
\begin{array}{l}
(X\ 1)\ \equiv\ 0\ 1\ 0\ 1\ 0\ 1\ 0\ 1 \\
(X\ 2)\ \equiv\ 0\ 0\ 1\ 1\ 0\ 0\ 1\ 1 \\
(X\ 3)\ \equiv\ 0\ 0\ 0\ 0\ 1\ 1\ 1\ 1 \\
\hline
(Z\ 1)\ \equiv\ 0\ 1\ 1\ 0\ 1\ 0\ 0\ 1 \\
(Z\ 2)\ \equiv\ 0\ 0\ 0\ 1\ 0\ 1\ 1\ 1
\end{array}
$$

Figure 1.4. Truth table of the full adder.

for configurations in the rows IY $\in$ {2,3}. The existence
function indicates four cases: IX $\in$ {3,5,6,7}, so
(Z 2) $\equiv$ (00010111).   The complete truth table (presented
horizontally, as by the APL programs) is shown in Fig. 1.4.

EXAMPLE 7.   Fig. 1.2d shows the existence function
of a NOR flip-flop with a reset terminal, X3.   The conventional diagram of this flip-flop is shown in Fig. 1.5.
The corresponding system is composed of the input subsystem
(X J); J = 1,2,3 and the output subsystem (Y K); K = 1,2.
The diagram in Fig. 1.5 was postulated as the only constraint of the system.   The equations of the circuit,
written in APL,

$$((\underline{Y}\ 2)\ =(\underline{Y}\ 1)\ \wedge\ (\underline{X}\ 1))\ \wedge\ ((\underline{Y}\ 1)\ =\ (\underline{Y}\ 2)\ \wedge\ (\underline{X}\ 2)\wedge\ (\underline{X}\ 3))$$

are satisfied for validity configurations corresponding
to windows where the existence function (Fig. 1.2d) is
true.   The circuit properties can be derived from that
function:

1.   It is clear that the existence function in
     Fig. 1.2d cannot be replaced by a truth table
     because not every column contains exactly one
     nonzero (see column IX = 0).   Thus, the circuit

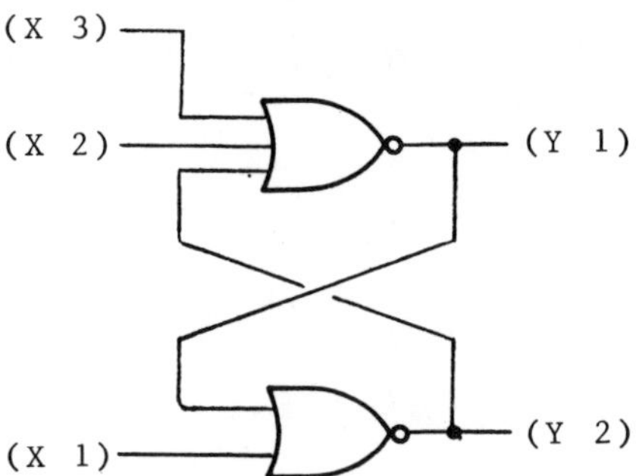

Figure 1.5.   NOR flip-flop with reset.

is not combinational but rather sequential
(containing feed-backs).

2. There are exactly nine steady states: Two for
    IX = 0 and one for each IX ∈ {1,2,3,4,5,6,7}.

3. When (X 3) = 1 (reset signal high), then
    IX ∈ {4,5,6,7}. All four existing validity
    configurations for that domain (right-hand
    half of the chart) have (Y 1) = 0 in common.
    That means that X3 ⇒ Y1 independently of
    anything else.

4. When (X J) = 0 for all J, then IX = 0 and the
    circuit can be in either of two steady states
    IX ∈ {1,2}, in which case (Y 2) ≠ (Y 1).

5. (Refer to Fig. 1.6) Starting with the steady
    state: IS = 16, the change of X1 alone (IX =
    0 → 1, column IX = 1) produces the unstable
    state: IS = 17, which goes over to the steady
    state: IS = 9.

    The change of X1 alone (IX = 1 → 0) now
    enforces the steady state: IS = 8.

IS	Y2	Y1	X3	X2	X1
16	1	0	0	0	0
17	1	0	0	0	1
9	0	1	0	0	1
9	0	1	0	0	1
8	0	1	0	0	0
8	0	1	0	0	0
10	0	1	0	1	0
18	1	0	0	1	0
18	1	0	0	1	0
16	1	0	0	0	0

Figure 1.6. State transitions
for flip-flop of Figure 1.5.

5.   (Contd.) The change of X2 alone (IX = 0 → 2,
      column IX = 2) produces the unstable state:
      IS = 10, which goes over to the stable state:
      IS = 18.

      A change of X2 alone (IX = 2 → 0) enforces the
      steady state: IS = 16.   Flip-flop transistion
      is thus illustrated.

The reader is now invited to go to Chapter 2, which
illustrates the use of the library program SYSTEM to
specify Boolean functions either by procedure SPACE
(existence function development) or by procedure TABLE
(producing a truth table of functions, either from their
sufficient functions or by listing).

The library program SYSTEM has two groups of procedures. The first prepares truth tables or existence
functions (discriminants) of a system subjected to a set
of constraints. The second group contains important
design procedures for the spacial treatment of Boolean
functions such as charting, minimization of $\Sigma\Pi$- and
$\Pi\Sigma$-forms, listing of prime implicants, and evaluation
of the Boolean difference. Some of the algorithms used
in the APL programs differ from those found in teaching
texts -- the triadic ordering of implicants, minimization
by extension of a $\Sigma\Pi$-form, and multiple-output design
optimization based on S-minimization of a mosaic Boolean
function are mentioned to name the most important.
Explanations of these algorithms will be given in the
second section of this text, and logical instruments
will be offered there as efficient means of teaching
the basic concepts.

The library program OPTIMA designs a multiple-output
two-level combinational circuit for a given set of incompletely specified Boolean functions (e.g., with "don't-
care"s). The result is printed out graphically.

The specifications are in the form of decimal equivalents:
Specified ONES and specified DON'T CARES. The printout
designates (X 1) with A, (X 2) with B, and so on. The
procedure is entered by calling DESIGN. Examples are
found in Chapter 4.

The library program BOOL solves systems of Boolean
equations of a general type by enumeration. The procedure is entered by calling BULL, and equations are entered
by calling FORMULA. The first literals of the alphabet
represent the Boolean constants (for instance: A, B, C,
D), and the literals that follow them in their natural
sequence represent the unknowns (for instance: E, F).
The values 0 (false) and 1 (true) may be used (alone!)
on the right-hand side of the formula only. The sum of
products form must be used on both sides of the formula.
Only two relations between the sides are accepted by the
programs, EQUIVALENCE (=) and IMPLICATION ($\rightarrow$) (the APL
right-arrow). Underlining may be used to represent
negation.

EXAMPLE 8. Examples of correctly-composed formulas
(it does not matter how many variables are constants and
how many are unknowns):

$$C\underline{A} + \underline{B}D = \underline{AB} + C\underline{D}A$$

$$A\underline{C}D + B = 0$$

$$E\underline{D} + \underline{E}D = 1$$

$$\underline{AB}D \rightarrow C\ \underline{E}$$

$$DC + C\underline{A} \rightarrow \underline{E} + B\ A$$

$$E\ D\ CB \rightarrow A + B + \underline{D}$$

Note that spaces within a product or around the signs are
acceptable. The sign "+" means OR; the sign "$\rightarrow$" means
"implies".

Chapter 2
Generation and Processing of Boolean Functions

2.1 INTRODUCTION

The program block, SYSTEM, contains programs for
(1) generating the existence function of a Boolean function from a set of constraints, (2) generating the truth
table of a Boolean function from constraints or from its
decimal equivalents, and (3) analyzing or processing a
Boolean function after it has been entered via either
of the first two procedures. This chapter describes and
illustrates the use of these programs.

2.2 EXISTENCE FUNCTION GENERATION

Fig. 2.1 shows the sequence of programs to be
executed. The sequence always starts with LOGIC, which
allows the user to type in the number of independent
variables: (X J), J = 1,2,...,NX. Function SPACE is
called to establish the number of dependent variables
(Y K), K = 1,2,...,NY in the system. Then, EQUATION is
called as many times as there are constraints postulated
for the system. Calling DISCRIMINANT causes the existence function of the system to be printed in the form
of a chart, where variables (X J) are represented horizontally and (Y K) are represented vertically. The
existence function depends on all variables of the system, and is true for each validity configuration.

$$(X_1, X_2, \ldots, X_{nx}, Y_1, Y_2, \ldots, Y_{ny}),$$

which satisfies all constraints of the system. The constraints form a system of Boolean equations, which can be solved by calling SOLVE after DISCRIMINANT has been formed. Each of the existing solutions is printed in the form

$$(Z\ K) = \text{FUNCTION OF } (X\ J)$$

for $K = 1, 2, \ldots, NZ$ with $NZ = NY$. The substitution of a solution in any equation (constraint) reduces that equation to an identity.

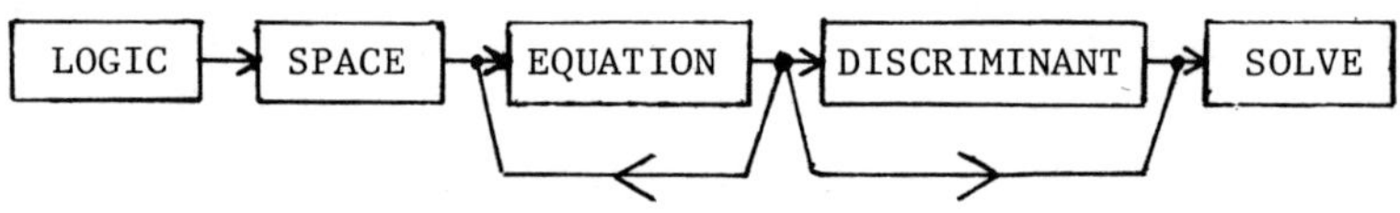

Figure 2.1.   Existence Function Generation.

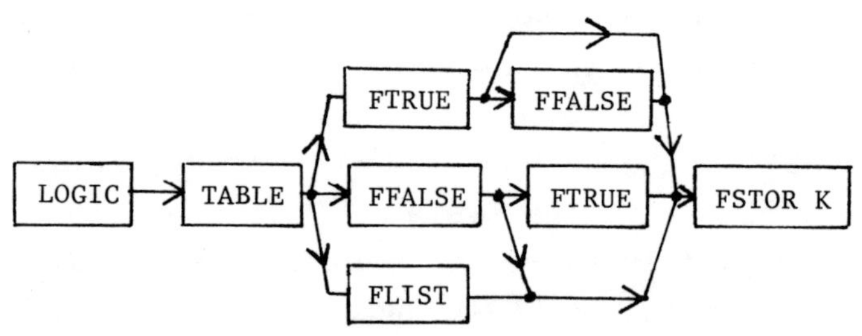

Figure 2.2.   Truth Table Generation.

2.3 TRUTH TABLE GENERATION

Fig. 2.2 shows the sequence of programs used to generate a truth table of Boolean functions (Z K), K = 1,2,...,NZ.

The sequence starts with LOGIC again (to set the number of independent variables (X J), J = 1,2,...,NX). Routine TABLE prepares the format of the truth table (by setting the number of functions (Z K), K = 1,2,...,NZ to be stored). The routines FTRUE and FFALSE develop Boolean functions from APL-defined constraints. The routine FLIST develops the Boolean functions from its decimal equivalents (the conventional approach).

2.4 PROCESSING OF BOOLEAN FUNCTIONS

To process Boolean functions for logic design, a number of routines are available. These are listed in Table 2.1.

ROUTINE	DESCRIPTION
F CHART WIDE	Prints the Marquand chart of the function F whose width is indicated by integer WIDE.
MINIMA F	Prints one of N-minimal $\Sigma\pi$-forms of the function F. Symbolism of the printout: A $\equiv$ (X 1), B $\equiv$ (X 2), etc.
PRIMIMPLICANT OF	Prints the sum of all prime implicants of the function OF in the algebraic form using the symbolism described for MINIMA.
DEGENERATION F	Looks for latent dependencies between variables in a system with existence function F.
K BULDIF F	Develops and prints the Boolean difference of the function F in relation to the variable (X K).
DECIMIN F	Produces decimal equivalents defining ONEs of a Boolean function F stored as a string of ZEROs, ONEs, and TWOs (DON'T CARES).
DECIDONT F	Produces decimal equivalents defining the DON'T CARES of a Boolean function F stored as defined for DECIMIN.

Table 2.1. Routines for
processing Boolean functions.

2.5 EXAMPLES

EXAMPLE 2.5.1. Develop the existence function of the full adder (see Fig. 1.3). Use its equation in APL given in formula 1.1.

Procedure:

Load program block SYSTEM.

Call ΔSYSTEM for explanations.

Begin by calling:

```
     LOGIC
NUMBER OF  X-VARIABLES:
□:
      3
NX= 3
SYMBOLS FOR X-VARIABLES:  (X J), (X̲ J); J=1   2   3
CALL:  TABLE, SPACE
```

To produce the existence function of this system, call:

```
     SPACE
NUMBER OF  Y-VARIABLES:
□:
      2
SYMBOLS FOR Y-VARIABLES:  (Y K), (Y̲ K); K= 1   2
NY= 2
XY-SPACE SYMBOL:   F[Y;X]
CALL: EQUATION
```

```
     EQUATION
WRITE THE EQUATION IN THE PRESCRIBED FORM:
F←F∧(ANY LOGICAL RELATION)
EXECUTE IT AND CALL: EQUATION, DISCRIMINANT, SOLVE
```

$$F \leftarrow F \wedge (((Y\ 1)+2\times(Y\ 2))=(X\ 3)+(X\ 2)+(X\ 1))$$

Note that at this moment the terminal is in the APL immediate-execution mode. The equation can be represented by <u>any</u> <u>number</u> of conditions executed immediately after one another:

$$F \leftarrow F \wedge (\text{Relation } 1)$$
$$F \leftarrow F \wedge (\text{Relation } 2)$$
$$F \leftarrow F \wedge (\text{Relation } 3)$$
$$\vdots \qquad \vdots$$
$$F \leftarrow F \wedge (\text{Relation } n)$$

Of course, each relation must be in the prescribed form. Routine DISCRIMINANT can be called at any time to check the evolution of the existence function.

When all relations (i.e., constraints of the system) have been introduced, the existence function is printed by calling DISCRIMINANT.

```
      DISCRIMINANT
HORIZONTAL SCALE:   (X J) FOR   J= 1   2   3
VERTICAL SCALE:   (Y K) FOR K= 1   2
  1 0 0 0 0 0 0 0
  0 1 1 0 1 0 0 0
  0 0 0 1 0 1 1 0
  0 0 0 0 0 0 0 1
```

(Compare this with 1.2c.)

The existence function always has the X-scale printed horizontally, and the Y-scale vertically. As with a Marquand chart, the X-coordinate $IX = (x3 \; x2 \; x1)_2 = 0,1,\ldots,7$, and the Y-coordinate (downward) $IY = (y2 \; y1)_2 = 0,1,2,3$. Each column contains exactly one non-zero. <u>For that reason</u>, we can develop the truth table of the full adder from its existence function. At the terminal, we do it by calling:

```
      SOLVE
NUMBER OF SOLUTIONS AFTER CONSTRAINTS: SOL= 1
DESIRED SOLUTION VECTOR:
□:
      1
SOLUTION NUMBER: 1
(Z 1) → 0   1   1   0   1   0   0   1       [1   2   4   7] ∪ ()
(Z 2) → 0   0   0   1   0   1   1   1       [3   5   6   7] ∪ ()
```

<u>Explanation</u>: The number of solutions is equal to the product of eight integers (for this example), one for each column. Each integer is equal to the count of non-zeros in the column. Here SOL = 1 (number of solutions); the solution vector has only one element, element 1 (first solution); and the resulting Boolean functions of the truth table are designated by (Z 1) for the output (Y 1) = 1 and (Z 2) for the output (Y 2) = 1. In the present case, functions do not contain DON'T CAREs (expressed by the integer 2 within the string on the left-hand side of the printout and by decimal equivalents in the parentheses at the right-hand side. To print the truth table, we call:

```
     FX
 0  1  1  0  1  0  0  1
 0  0  0  1  0  1  1  1
```

To understand the reason for using the symbol (Z 1) for the output Boolean function at the terminal (Y 1), let us call:

```
     (Y 1)
 0 0 0 0 0 0 0 0 1 1 1 1 1 1 1 1 1 0 0 0 0 0 0 0 0 0 1 1 1 1
 1 1 1 1
```

```
     (Y 2)
 0 0 0 0 0 0 0 0 0 0 0 0 0 0 0 0 0 1 1 1 1 1 1 1 1 1 1 1 1 1
 1 1 1 1
```

The N-minimal $\Sigma\pi$-form of the functions (Z K) of the truth table can be obtained immediately by the procedure MINIMA.

```
     MINIMA ( Z 1 )
ABC + ABC + ABC + ABC
CRITICAL SET: 7   4   2   1

     MINIMA ( Z 2 )
BC + AC + AB
CRITICAL SET: 6   5   3

     MINIMA ( Z 2 )
AB + AC + BC
CRITICAL SET: 4   2   1
```

```
     MINIMA (Z 1)
ABC + ABC + ABC + ABC
CRITICAL SET: 6   5   3   0
```

The minimization results suggest immediately

$$(\underline{AB} + \underline{AC} + \underline{BC})(A + B + C) = A\underline{BC} + \underline{A}B\underline{C} + \underline{AB}C$$

so that

$$(Z\ 1) \equiv (\underline{Z}\ 2)(A + B + C) + ABC$$

(The preceding symbolism is Boolean: Multiplication
stands for AND, and addition (+) stands for OR. Note
that the $\Sigma\pi$-forms are printed by the terminal using this
same symbolism.)

EXAMPLE 2.5.2. Full adder analysis based on the
truth table of threshold functions of its input (Fig. 1.3
again). The sequence of procedures is illustrated in
Fig. 2.3.

```
     LOGIC
NUMBER OF  X-VARIABLES:
□:
     3
NX= 3
SYMBOLS FOR X-VARIABLES: (X J), (X J); J=1   2   3
CALL:  TABLE, SPACE
```

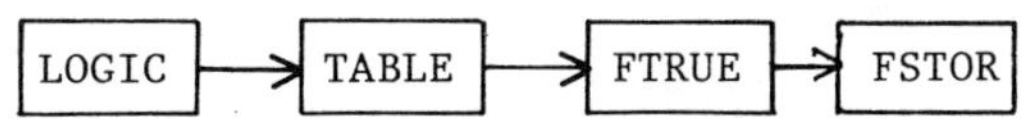

Figure 2.3. Program module sequence for
truth table of threshold functions.

```
     TABLE
 NUMBER OF FUNCTIONS:
☐:
     5
TABLE IS READY FOR FUNCTIONS  (Z K)  WITH  K= 1   2   3
  4   5
CALL OFFERINGS: FTRUE, FFALSE, FLIST

     FTRUE
WRITE A SUFFICIENT CONDITION OF  F  IN THE PRESCRIBED
FORM:
   F←(ANY LOGICAL APL-MEANINGFUL RELATION WITH VARIABLES
(X J), (X J))
EXECUTE IT AND CALL
EITHER:  FSTOR K  (WHERE  K  IS THE INDEX OF THE FUNCTION
(Z K)
     OR:  FFALSE


   F←0<(X 1)+(X 2)+(X 3)


     FSTOR 1
CALL:  FTRUE,  FFALSE,  FLIST  TO DEFINE THE NEXT
FUNCTION   (F K)
WITH K=2
```

Again, a fixed form is prescribed for the condition
definition. The variable **F** represents the Boolean function defined by the APL expression. The procedure FSTOR 1
will store it as the first row of the truth table. As a
result of the response to program TABLE, there are five
rows in the table ready to be filled with threshold functions.

The next threshold function is entered by:

```
     FTRUE
WRITE A SUFFICIENT CONDITION OF  F  IN THE PRESCRIBED
FORM:
   F←(ANY LOGICAL APL-MEANINGFUL RELATION WITH VARIABLES
(X J), (X J))
EXECUTE IT AND CALL
EITHER:  FSTOR K  (WHERE  K  IS THE INDEX OF THE FUNCTION
(Z K)
     OR:  FFALSE


   F←1<(X 1)+(X 2) + (X 3)
```

```
      FSTOR 2
 TRUTH TABLE IS READY:
MAY CALL:  FTRUE, FFALSE,  FLIST FOR  K= 3 ≤5
MAY EXECUTE  FX  TO PRINT THE TABLE
```

Two rows of the truth table have now been filled. A third threshold function will be placed in the third row.

```
      FTRUE
WRITE A SUFFICIENT CONDITION OF  F  IN THE PRESCRIBED
FORM:
   F←(ANY LOGICAL APL-MEANINGFUL RELATION WITH VARIABLES
(X J), (X̲ J))
EXECUTE IT AND CALL
EITHER:  FSTOR K  (WERE  K  IS THE INDEX OF THE FUNCTION
(Z K)
    OR:  FFALSE

   F←2<(X 1)+(X 2)+(X 3)

      FSTOR 3
 TRUTH TABLE IS READY:
MAY CALL:  FTRUE, FFALSE,  FLIST FOR  K= 4 ≤5
MAY EXECUTE  FX  TO PRINT THE TABLE
```

To check the progress of function generation, let us print the state of the truth table:

```
     FX
 0 1 1 1 1 1 1 1
 0 0 0 1 0 1 1 1
 0 0 0 0 0 0 0 1
 0 0 0 0 0 0 0 0
 0 0 0 0 0 0 0 0
```

One of the adder's output functions (designated in Fig. 1.3 by Y1) will be placed in the next (fourth) row of the truth table:

```
      FTRUE
WRITE A SUFFICIENT CONDITION OF  F  IN THE PRESCRIBED
FORM:
   F←(ANY LOGICAL APL-MEANINGFUL RELATION WITH VARIABLES
(X J), (X̲ J))
EXECUTE IT AND CALL
EITHER:  FSTOR K  (WHERE  K  IS THE INDEX OF THE FUNCTION
(Z K)
    OR:  FFALSE

   F←2|(X 1)+(X 2)+(X 3)
```

```
     FSTOR 4
 TRUTH TABLE IS READY:
MAY CALL:  FTRUE, FFALSE,  FLIST FOR  K= 5  ≤5
MAY EXECUTE  FX  TO PRINT THE TABLE
```

The truth table printout now appears as:

```
     FX
 0  1  1  1  1  1  1  1
 0  0  0  1  0  1  1  1
 0  0  0  0  0  0  0  1
 0  1  1  0  1  0  0  1
 0  0  0  0  0  0  0  0
```

It is clear that the function of the fourth row can
be obtained by Boolean algebraic combination of the functions of rows 1, 2, and 3. To prove this, let us put the
function $F \leftarrow (Z\ 3) \vee ((\underline{Z}\ 2) \wedge (Z\ 1))$ into row 5 of the
table:

```
     FTRUE
WRITE A SUFFICIENT CONDITION OF  F  IN THE PRESCRIBED
FORM:
   F←(ANY LOGICAL APL-MEANINGFUL RELATION WITH VARIABLES
(X J), (X J))
EXECUTE IT AND CALL
EITHER:  FSTOR K  (WHERE  K  IS THE INDEX OF THE FUNCTION
(Z K)
    OR:  FFALSE

     F←(Z 3)∨(Z 2)∧(Z 1)

     FSTOR 5
 TRUTH TABLE IS READY:
MAY CALL:  FTRUE, FFALSE,  FLIST FOR  K= 6  ≤5
MAY EXECUTE  FX  TO PRINT THE TABLE
```

```
     FX
 0  1  1  1  1  1  1  1
 0  0  0  1  0  1  1  1
 0  0  0  0  0  0  0  1
 0  1  1  0  1  0  0  1
 0  1  1  0  1  0  0  1
```

Note that the second output function of the adder
(designated by $Y2$ in Fig. 1.3) happens to be the threshold function of the inputs stored in the second row of
the table. To complete the analysis, the functions are

N-minimized:

MINIMA (Z 1)
C + B + A
CRITICAL SET: 4   2   1

MINIMA (Z 3)
ABC
CRITICAL SET: 7

MINIMA (Z̲ 2)
A̲B̲ + A̲C̲ + B̲C̲
CRITICAL SET: 4   2   1

MINIMA (Z 2)
BC + AC + AB
CRITICAL SET: 6   5   3

By inspection of the truth table and by using the results of the minimizations, we get

$$(Z\ 5) = (Z\ 3) \lor ((\underline{Z}\ 2) \land (Z\ 1)) \qquad (2.1)$$
$$= ABC + (\underline{Z}\ 2)(A + B + C)$$
$$= ABC + DA + DB + DC$$
$$\text{where } D = (\underline{Z}\ 2)$$
$$= \sim(BC + AC + AB).$$

The resulting equations are implemented in the two-bit binary full adder (for instance, SC7482).

EXAMPLE 2.5.3.   Develop the existence function of a given sequential circuit, the S-R NOR flip-flop (Fig. 1.5, Example 1.7).   Because of the feedback loop, sequential circuit behavior is expected.   The truth table generation procedure is inadequate for expressing this behavior.   The proper procedure sequence is shown in Fig. 2.4.

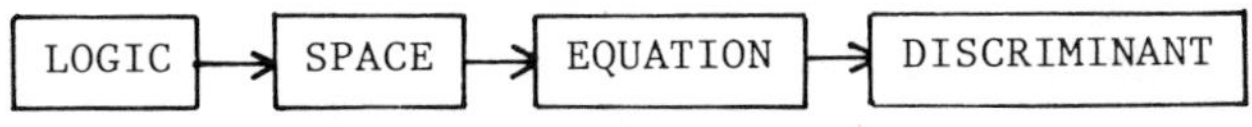

Figure 2.4.   Program module sequence for the existence function of a sequential circuit.

To begin, we call:

```
        LOGIC
NUMBER OF  X-VARIABLES:
□:
        3
NX= 3
SYMBOLS FOR X-VARIABLES: (X J), (X̲ J); J=1   2   3
CALL:  TABLE, SPACE

        SPACE
NUMBER OF  Y-VARIABLES:
□:
        2
SYMBOLS FOR  Y-VARIABLES: (Y K), (Y̲ K); K= 1   2
NY= 2
XY-SPACE SYMBOL:  F[Y;X]
CALL: EQUATION

        EQUATION
WRITE THE EQUATION IN THE PRESCRIBED FORM:
F←F∧(ANY LOGICAL RELATION)
EXECUTE IT AND CALL: EQUATION, DISCRIMINANT, SOLVE
```

The note given in Example 2.5.1 shows that it is unnecessary to condense all constraints into a single APL expression, which for this circuit would be

$$F \leftarrow F \wedge ((Y\ 2) = (\underline{Y}\ 1) \wedge (\underline{X}\ 1))$$
$$\wedge ((Y\ 1) = (\underline{Y}\ 2) \wedge (\underline{X}\ 2) \wedge (\underline{X}\ 3))$$

Clearer expressions are obtained by introducing and executing the simultaneous constraints one after the other, as it is done here:

```
F←F∧(Y 2)=(Y̲ 1)∧(X̲ 1)
F←F∧(Y 1)=(Y̲ 2)∧(X̲ 2)∧(X̲ 3)
```

Again, remember that APL executes these expressions as soon as they are entered. The existence function is now printed by calling:

```
        DISCRIMINANT
HORIZONTAL SCALE: (X J) FOR  J= 1   2   3
VERTICAL SCALE: (Y K) FOR K= 1   2
 0 0 0 1 0 1 0 1
 1 1 0 0 0 0 0 0
 1 0 1 0 1 0 1 0
 0 0 0 0 0 0 0 0
```

Each non-zero of the existence function represents a steady state of the circuit. The present circuit has nine steady states.

Note that this existence function agrees with Fig. 1.2d (Example 7, Chapter 1).

EXAMPLE 2.5.4. Design a combinational network specified only by its truth table. The table is given by "decimal equivalents". In that case, the procedure sequence of Fig. 2.5 is indicated. The given function has six independent variables.

```
      LOGIC
NUMBER OF X-VARIABLES:
□:
      6
NX= 6
SYMBOLS FOR X-VARIABLES: (X J); (X̲ J); J=1   2   3   4   5   6
CALL:  TABLE, SPACE

      TABLE
 NUMBER OF FUNCTIONS:
□:
      1
TABLE IS READY FOR FUNCTIONS  (Z K)   WITH K= 1
CALL OFFERINGS: FTRUE, FFALSE, FLIST
```

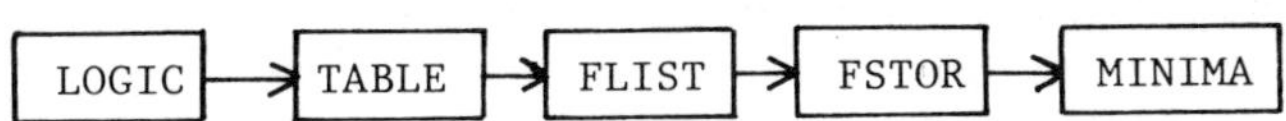

Figure 2.5. Program module sequence for
combinatorial design via truth table.

Here, we will select FLIST:

FLIST
DECIMAL EQUIVALENTS OF ONES (AT LEAST ONE ITEM):
☐:
      1 8 14 17 18 19 20 22 24 27 34 36 38 43 46 52 56
59 60 61 63
DECIMAL EQUIVALENTS OF DONT CARES:
☐:
      2 3 4 5 6 7 10 11 12 23 25 26 28 33 35 37 41 45 49
53 54 55 57 58 62
CALL: FSTOR K (WHERE K IS WELL SPECIFIED)

FSTOR 1
CALL: FTRUE, FFALSE, FLIST TO DEFINE THE NEXT
FUNCTION   (F K)
WITH K=2

The given Boolean function is now stored as (Z 1), and we want to inspect its Marquand chart. To print that chart, we call:

(Z 1) CHART 8
HORIZONTAL SCALE: (X J) FOR J= 1   2   3
VERTICAL SCALE: (X J) FOR J= 4   5   6

0	1	2	2	2	2	2	2
1	0	2	2	2	0	1	0
0	1	1	1	1	0	1	2
1	2	2	1	2	0	0	0
0	2	1	2	1	2	1	0
0	2	0	1	0	2	1	0
0	2	0	0	1	2	2	2
1	2	2	1	1	1	2	1

Note that the procedure is called with two arguments: On the left side of the procedure name is the symbol of the function to be charted; on the right side is the number of columns of the chart, which must be a power of two (for triadic charts, a power of three).

The chart is filled with integers from the set {0,1,2} with the following meanings: 0 means FALSE, 1 means TRUE, and 2 means UNSPECIFIED (DON'T CARE).

To design a two-level AND-OR (two-level NAND) combinational network, we first perform the N-minimization of (Z 1):

MIMINA (Z 1)
ABCE + BCDE + ACDF + ACD + DEF + BCF + ACDF + ACEF
CRITICAL SET: 46   34   8   36   63   18   17   43

That calls for eight gates at the first level and an 8-input fan-in at the second level. The first-level gates need 29 inputs. Inversion of B is not necessary. The maximum number of inputs per gate is 4.

To design a two-level OR-AND (two-level NOR) combinational circuit for the given function, we N-minimize the complement (Z̲ 1) of the function (Z 1) and develop an N-minimal ΠΣ-form by inverting the form obtained for (Z̲ 1):

MINIMA (Z̲ 1)
ABCD + ACE + BDEF + ABCD + CDEF + CEF + ADEF
CRITICAL SET: 16   47   44   42   50   31   9

That calls for seven gates at the first level and a 7-input fan-in at the second level. The first-level gates need only 26 inputs. It is clear that this solution is simpler. Note that

$$(Z\ 1) = (A+B+C+D)(A+C+E)(B+D+E+F)$$
$$(A+B+C+D)(C+D+E+F)(C+E+F)(A+D+E+F)$$

The low value of the execution time connected with the MINIMA procedure is due to the fact that the ΣΠ-extension algorithm avoids the listing of all prime implicants of the given function, which is time-consuming. To illustrate this point, the sum of <u>all</u> prime implicants is printed out by calling

PRIMIMPLICANT (Z 1)
ALGEBRAIC FORM OF THE COMPLEMENTARY FUNCTION:
BCDEF + ABDEF   + ACDEF + ABCEF + BCDEF + ABCEF + ABDEF +
ABCD + ACDF
CRITICAL SET: 51   44   42   39   30   21   9   16   15
SUM OF ALL PRIME IMPLICANTS:
ABCD + ACD + ABCD + ABCE + ABCE + ABDE + CDE + BCDE +
ABCE + BCDF + CEF + DEF + ACEF + ABCF + BCF + ABDF +
ACDF + ABCE + ABDE + ABDE + BCDE + ACDE + BCDE + ABF +
ABCF + ACDF + BDF + ACDF + ACEF + ABEF + ACEF + ADEF +
ADEF + CDEF

The function (Z 1) has 34 prime implicants. Only eight of them were used to form an N-minimal $\Sigma\Pi$-form. This is wasteful. The procedure MINIMA developed prime implicants covering points belonging to a CRITICAL SET: $\{46,34,8,36,63,18,17,43\}$, whose elements are MUTUALLY TERM EXCLUSIVE (that is, no term implicant of the given function exists that covers any pair of elements of that set) and which has the property to include the maximum number of mutually term exclusive elements.

Notice that the algebraic form of the complementary function covers all zeros of the given function without being N-minimal. The N-minimal $\Sigma\Pi$-form of ($\underline{Z}$ 1) is much simpler due to DON'T CAREs, which are not considered for the algebraic form of the printout.

EXAMPLE 2.5.5. When the function is very simple and when all we want to do is minimize its algebraic form, it is possible to develop its APL representation without beginning with LOGIC. For instance, a Boolean function FX is specified by decimal equivalents as follows: (FX = 1) is specified for the set of decimal equivalents $\{3,5,6\}$, and FX is unspecified for $\{1,2,4,7\}$. The APL vector will be a string of eight integers (it is obvious that the function has three independent variables) taken from the set $\{0,1,2\}$: 0 means FALSE, 1 means TRUE, and 2 means UNSPECIFIED (DON'T CARE). At the APL terminal, the function can be developed by:

```
        FX ← 0   2   2   1   2   1   1   2
  FALSE →  0
   TRUE →               3       5   6
UNSPECIFIED →    1   2       4           7
```

The function can thus be minimized immediately as follows:

```
    FX←0 2 2 1 2 1 1 2
```

GENERATION AND PROCESSING OF BOOLEAN FUNCTIONS

```
      MINIMA FX
CYCLE DISSOLUTION
C + B
CRITICAL SET:  6   3
```

The given function FX was "cyclic". In that case,
the procedure MINIMA dissolves the cycle by relaxing the
condition leading to the minimal count of literals in the
resulting $\Sigma\Pi$-form. The number of terms of that form
remains minimal. However, the critical set printout can
be distorted. For instance, in the present case it can
be shown that no critical set exists that has two elements. The set {6,3} is not "critical" because the term
B covers both elements 6 and 3. On the other hand, all
N minimal $\Sigma\Pi$-forms of FX: C+B, B+A, A+B have two terms,
more than the number of elements in critical sets {6},
{3}, and {5}, possessing one element only. The function
FX is labeled, "abnormal" because the N-minimal $\Sigma\Pi$-form
always has more terms than the count of elements in its
critical set.

To meet an abnormal Boolean function in actual computer application is highly improbable. Cyclic functions
usually happen to be normal. For instance, the function
treated in the following is cyclic, but normal:

```
      LOGIC
NUMBER OF X-VARIABLES:
□:
      4
NX= 4
SYMBOLS FOR X-VARIABLES:  (X J),  (X̲ J);  J=1   2   3   4
CALL:  TABLE, SPACE

      TABLE
 NUMBER OF FUNCTIONS:
□:
      1
TABLE IS READY FOR FUNCTIONS  (Z K)  WITH  K= 1
CALL OFFERINGS: FTRUE, FFALSE, FLIST
```

```
      FLIST
 DECIMAL EQUIVALENTS OF ONES (AT LEAST ONE ITEM):
 □:
       0  1  3  4  6  7  9  10  11  12  13  14
 DECIMAL EQUIVALENTS OF  DONT CARES:
 □:
       ι0
 CALL:  FSTOR   K (WHERE K IS WELL SPECIFIED)
       FSTOR 1
 CALL:  FTRUE,  FFALSE,  FLIST  TO DEFINE THE NEXT
 FUNCTION    (F K)
  WITH  K=2

       (Z 1) CHART 4
 HORIZONTAL SCALE: (X J) FOR   J= 1   2
   VERTICAL SCALE: REMAINING VARIABLES.
  1  1  0  1
  1  0  1  1
  0  1  1  1
  1  1  1  0

       MINIMA (Z 1)
 CYCLE DISSOLUTION
 BCD + AC + ABD + BCD + ABD
 CRITICAL SET: 13   9   10   7   0
```

Note again that "cycle dissolution" means that the critical set may be faulty. Here, minterms 13 and 9 are mutually term <u>inclusive</u> (both being covered by ABD), which is against the ruling that each pair of elements in the critical set must be mutually term <u>exclusive</u>. The resulting form is always correctly N-minimal. The minimal count of literals is not warranted. To get the best N-minimal $\Sigma\Pi$-form of cyclic functions, use the program block called OPTIMA.

EXAMPLE 2.5.6. The procedure X NEWORDER F is used to interchange the indexing of variables (i.e., to change the charting pattern of F). This procedure is used as a subroutine in BULDIF generating the Boolean difference of F. The dummy variable X is used to express the reordering law; it has the form of a vector composed from the elements of the set {1,2,3,...} containing one element for each variable (X J), J = 1,2,3,... of the function F.

Note that these variables are represented in the MINIMA printout as A, B, C, ... For instance,

$$2\ 3\ 4\ 1\quad NEWORDER\quad F$$

is well formed when F is a function of four variables (X J), J = 1,2,3,4 (represented by A, B, C, and D in the MINIMA printouts). To describe the reordering law, the following sketch is useful:

Original order	1	2	3	4 = J old	A	B	C	D
X-symbol		2	3	4	1		2	3	4	1
New order		4	1	2	3 = J new	D	A	B	C

The variable (J 1) of the new order represents the variable (J 4) of the old. In the printout, printing of A will be replaced by the printing of D (note that new D stands below the old A in the preceding sketch).

To illustrate, start with

$$F1\leftarrow 1\ 0\ 1\ 0\ 1\ 0\ 1\ 0\ 0\ 1\ 0\ 1\ 1\ 1\ 0\ 1$$

and do

```
     F1 CHART 4
HORIZONTAL SCALE: (X J) FOR  J= 1   2
  VERTICAL SCALE: (X J) FOR  J= 3   4
 1 0 1 0
 1 0 1 0
 0 1 0 1
 1 1 0 1
```

to print the Marquand chart of F1. In this chart, the horizontal variables are A and B; the vertical variables are C and D. The MINIMA printout gives:

```
     MINIMA F1
AD + ABC + AD
CRITICAL SET: 15   12   6
```

The reordering procedure, called by

$$F4\leftarrow 2\ 3\ 4\ 1\quad NEWORDER\quad F1$$

produces the function F4, whose variables are reordered.

To show the result of the reordering operation, call:

```
     MINIMA F4
CD + ABD + CD
CRITICAL SET: 15   6   3
```

Note that the printout of MINIMA F1 was transformed
into that of MINIMA F4 according to the sketch discussed
previously: AD transforms into DC, ABC into DAB, and AD
into DC.

Second illustration:

```
 F4←2 1 3 4 NEWORDER F1
```

This time, the variables A and B were mutually interchanged, while C and D were left unchanged. The Marquand
chart after this reordering is produced by

```
     F4 CHART 4
HORIZONTAL SCALE: (X J) FOR   J= 1   2
  VERTICAL SCALE: (X J) FOR   J= 3   4
 1  1  0  0
 1  1  0  0
 0  0  1  1
 1  0  1  1
```

Comparison of this chart with the chart of the original function F1 shown previously shows that the column
corresponding to BA was interchanged with the column
belonging to BA. Finally, note the result of

```
     MINIMA F4
BD + ABC + BD
CRITICAL SET: 15   12   5
```

EXAMPLE 2.5.7. Find the Boolean difference of a
given function F1 (from example 2.5.6) in relation to
each of its variables (X K), K = 1,2,3,4. A procedure
is available to solve this problem. By calling:

$$R \leftarrow K \ BULDIF \ F \ ,$$

a function of _all_ variables is formed. This function is
true for a given configuration of all variables' validi-

ties if and only if the value of the function F varies
with the variation of the variable (X K).

   To illustrate, let us call:

   F4←1 BULDIF F1

and express F4 (the difference for (X 1)) algebraically
by calling:

   MINIMA F4
B + C̲ + D̲
CRITICAL SET: 15   9   5

   In the sense of the definition, the function F1
must change its value if and only if

$$F4 = 1 = B + C̲ + D̲.$$

To check it, we recall that

$$F1 = AD + A̲B̲C + A̲D̲.$$

   Test 1: B = 1, C,D = (any) => F4 = 1.
            Then F1 = AD + A̲D̲ ≡ (A = D).
            Conclusion: F1 must change with
            A for any D.

   Test 2: D = 0, B,C = (any) => F4 = 1.
            Then F1 = A̲B̲C + A̲ ≡ A̲, which
            changes with A.

   Test 3: B̲CD = 1 => F4 = 0.
            Then B = C̲ = D̲ = 0,
             and F1 = A + A̲ = 1
            so that F1 does not change with A.

   Another illustration:

   F4←4 BULDIF F1

   MINIMA F4
B + A + C̲
CRITICAL SET: 14   13   8

```
      F4  CHART  4
HORIZONTAL  SCALE:  (X  J)  FOR    J = 1   2
   VERTICAL  SCALE:  (X  J)  FOR    J = 3   4
     1   1   1   1
     0   1   1   1
     1   1   1   1
     0   1   1   1
```

The chart shows that the value of F1 changes with (X 4) (or D) in every window of the chart except two (those filled with zeros).

Chapter 3
The APL Program "SYSTEM"

This chapter is a collection of program modules, called functions in APL. Together they comprise a program set known as SYSTEM. These modules are the ones referenced throughout Chapter 2.

The serious student of APL will wish to study these in depth. The reader who is only interested in their application to circuit design may skip this chapter.

These modules have been developed by Dr. Svoboda over the past several years and are successfully used in his popular Logical Circuits Laboratory course at the University of California, Los Angeles. The user, either student or scientist, needs only a rudimentary introduction to APL to be able to apply SYSTEM to design problems.

SYSTEM is used to transform problem specifications into a set of Boolean functions defined by a truth table. It is the basic starting program for study as it will provide the existence function of the system of Boolean functions and is the program which provides for minimization and logical relational analysis.

```
      ∇ R←DECIDONT F
[1]     R←¯1+(F=2)/ιρF
      ∇


      ∇ R←DECIMIN F
[1]     R←¯1+(F=1)/ιρF
      ∇
```

```
     ∇ R←K BUILDIF F;A;B;C;G;D;H
[1]      →(B≠⌊B←2⊛A←0.5×A←ρF←2-F)/0
[2]      D←⌽(2,A)ρ2×ιA
[3]      C←ιB←B+1
[4]      C[K,B]←B,K
[5]      G←C NEWORDER G←A⌽G←C NEWORDER F
[6]      G←G×F
[7]      R←(G=2)+2×G=0
[8]      H←,⌽((2*K-1),2*1+B-K)ρD
[9]      R←(R×R≠2)+H×R=2
[10]     'DONT CARES DESIGNATED WITH EQUAL NUMBERS'
[11]     '(LARGER THAN ONE) MUST BE GIVEN EQUAL TRUTH-VALUES.
     ∇


     ∇ YX←F CHART WIDE;XH;XV;MX;MY
[1]     XV←(ρF)÷WIDE
[2]     XH←WIDE
[3]     YX←(XV,XH)ρF
[4]     MX←((0=2|XH)×(2⊛XH))+(0=3|XH)×(3⊛XH)
[5]     MY←(ρF)÷MX
[6]     'HORIZONTAL SCALE: (X J) FOR  J= ';ιMX
[7]     '  VERTICAL SCALE: REMAINING VARIABLES.'
     ∇

     ∇ R←COMB N;BMB;P;U;CMB
[1]     BMB←R←2*(¯1+ιN)
[2]     P←1
[3]  H1:CMB←ι0
[4]     U←P
[5]  H2:CMB←CMB,((BMB<2*U)/BMB)+2*U
[6]     →(N>U←U+1)/H2
[7]     R←R,CMB
[8]     BMB←CMB
[9]     →(N>P←P+1)/H1
     ∇

     ∇ DEGENERATION F;NX;NNX;L1;F0;F1;F2;F3;F4;F5;F6;F7;F8;
continue on the same line..   LIM;U;R;MX;F9;V;K
[1]     F3←'ABCDEFGHJKLMN'
[2]     →(0=F0←+/(F=1))/0
[3]     L1←F0←⌈2⊛F0
[4]     →(F0<NX←2⊛NNX←ρF)/D1
[5]     'NO DEGEN.'
[6]     →0
[7]  D1:L1←NX-F0
[8]     F5←ιNX
[9]     F1←COMB NX
[10]    F2←NXρ2
[11]    LIM←ρF1
[12] D7:U←1
[13] D2:→(L1=+/1=F4←F2⊤F1[U])/D3
```

```
[14]  D8:→(LIM≥U←U+1)/D2
[15]   →D4
[16]  D3:F6←F4/F5
[17]   F7←((~F5∈F6)/F5),(F5∈F6)/F5
[18]   R←F7 NEWORDER(F=1)
[19]   MX←((2*ρF6),2*NX-ρF6)ρR
[20]   F8←+/MX
[21]   →(0=+/F8>1)/D5
[22]   →D8
[23]  D5:'VARIABLES (X ';F6;') ARE FUNCTIONS OF ALL
                 REMAINING VARIABLES'
[24]   'DISCRIMINANT'
[25]   MX
[26]   V←K←1
[27]  D6:F9←NNXρ(Vρ0),Vρ1
[28]   F9←⍉((2*NX-L1),2*L1)ρF9
[29]   F9←MX×F9
[30]   F9←+/F9
[31]   '(X ';F6[K[;') → ';F9
[32]   V←2×V
[33]   →(L1≥K←K+1)/D6
[34]   →D8
[35]  D4:→(0<L1←L1-1)/D7
[36]   →0
    ∇

    ∇ DISCRIMINANT
[1]   ⍝ EXISTENCE FUNCTION CHART
[2]    F CHART NNX
[3]    R←(NX,(0.5×NNX))ρ0
    ∇




    ∇ EQUATION
[1]   ⍝ CONSTRAINT INSERTION
[2]    'WRITE THE EQUATION IN THE PRESCRIBED FORM:'
[3]    'F←F∧(ANY LOGICAL RELATION)'
[4]    'EXECUTE IT AND CALL:  EQUATION, DISCRIMINANT, SOLVE'
    ∇




    ∇ FFALSE
[1]    TS←CM←5
[2]    ' WRITE A SUFFICIENT CONDITION OF  F  IN THE FORM:'
[3]    '       F←',TEX1
[4]    ' EXECUTE IT AND CALL'
[5]    ' EITHER:  FSTOR K  (WHERE K IS THE INDEX OF THE
      FUNCTION  (Y K) )'
[6]    '      OR:  FTRUE'
    ∇
```

```
      ∇ FLIST;ONE;DNT
[1]    ⍝ FUNCTION DEFINITION BY LISTING DECIMAL
       EQUIVALENTS
[2]    ' DECIMAL EQUIVALENTS OF ONES (AT LEAST ONE
       ITEM):'
[3]    ONE←,ONE←⎕
[4]    →(0=ρONE)/0
[5]    F←(ιNNX)∊1+ONE
[6]    ' DECIMAL EQUIVALENTS OF DONT CARES:'
[7]    DNT←,DNT←⎕
[8]    F←~(ιNNX)∊(1+ONE,DNT)
[9]    ' CALL:  FSTOR  K (WHERE K IS WELL SPECIFIED)'
      ∇
```

```
      ∇ R←NX FORM TR;TST;V;H;U;LIM
[1]    R←ι0
[2]    V←1
[3]    LIM←(ρ,TR)÷NX
[4]   P1:H←TR[V;]
[5]    U←1
[6]   P2:→(0=H[U])/P3
[7]    R←R,G1[(2|H[U])+2×U]
[8]   P3:→(NX≥U←U+1)/P2
[9]    →(LIM<V←V+1)/0
[10]   P←R,' + '
[11]   →P1
      ∇
```

```
      ∇ FSTOR K
[1]    ⍝ TRUTH TABLE GENERATION
[2]    →(4≠TR+TS)/S1
[3]    FX[K;]←F
[4]    →S2
[5]   S1:→(5≠TR+TS)/S3
[6]    FX[K;]←~F
[7]    →S2
[8]   S3:→(0=+/F×F)/S4
[9]    ' CONTRADICTION, TRY AGAIN BY RECALLING  FTRUE  OR
       FFALSE  (RESP.: FSET )'
[10]   →0
[11]  S4:FX[K;]←F+2×~F+F
[12]  S2:TR←TS←0
[13]   →(K≥NY)/S5
```

```
[14]    ' CALL:   FTRUE,   FFALSE,   FLIST   TO DEFINE THE
        NEXT FUNCTION   (F K)'
[15]    ' WITH   K=';K+1
[16]    →0
[17] S5:' TRUTH TABLE IS READY:'
[18]    'MAY EXECUTE   FX   TO PRINT THE TABLE'
     ∇


     ∇ FTRUE
[1]     TR←CM←4
[2]     'WRITE A SUFFICIENT CONDITION OF  F   IN THE
        PRESCRIBED FORM:'
[3]     '    F←',TEX1
[4]     'EXECUTE IT AND CALL'
[5]     'EITHER:  FSTOR K  (WHERE   K   IS THE INDEX OF
        THE FUNCTION (Z K) )'
[6]     '     OR:   FFALSE'
     ∇


     ∇ LOGIC
[1]     G1←' AABBCCDDEEFFGGHHJJKKLLMMNN'
[2]     'NUMBER OF   X←VARIABLES:'
[3]     NNX←2*NX←□
[4]     'NX= ';NX
[5]     'SYMBOLS FOR X-VARIABLES: (X J), (X J); J=';ιNX
[6]     BINX←NXρ2
[7]     XX←⊖BINXτ(¯1+ιNNX)
[8]     'CALL:  TABLE, SPACE'
     ∇


     ∇ R←MIN I
[1]     R←2-XX[;I+1]
     ∇
```

```
        ∇ MINIMA F;LIM;NX;G;NNX;TRN;S;R;ZE;H;FND;RZZ;NW;U;
          V;IND;NM;RES;RXX;TSP;TST;SUT;TSF;SUF;MEZ;EFF;
          BINZ;DIF;TRY;FSP
[1]       LIM←2+NX←2⊛NNX←ρF
[2]       AMB←COMB NX
[3]       R←CRI←FND←ι0
[4]       NZ←+/ZE←F=DIS←0
[5]       S←1
[6]    A0:H←((2*NX-S),2*S)ρZE
[7]       FND←FND,,H∨(2*S-1)⌽H
[8]       →(NX≥S←S+1)/A0
[9]       FND←(NX,NNX)ρFND
[10]      G←(LIM×F>1)+G+(+⌿FND)×G←F=1
[11]      FND←(2×FND)+⊖(NXρ2)⊤(⁻1+ιNNX)
[12]      RZZ←2-2|ZE/FND
[13]      FSP←ZE/ιNNX
[14]   A4:→(0=NW←⌈/(G<LIM)/G)/A8
[15]      NM←+/RES←(G>0)∧G<LIM
[16]      RXX←2-2|RES/FND
[17]      TSP←RES/ιNNX
[18]   A3:→(0=U←ρ,IND←(G∈NW)/ιNNX)/A6
[19]   A2:TRY←TRM←(NIM>1)×2-2|NIM←,FND[;IND[U]]
[20]      TST←⍉(NM,NX)ρTRM
[21]      SUT←,TSP[((+⌿3=RXX+TST)=0)/ιNM]
[22]      TST←⍉(NZ,NX)ρTRM
[23]      SUF←,FSP[((+⌿3=RZZ+TST)=0)/ιNZ]
[24]      →(0=ρSUF)/A1
[25]      TRN←(2-2|,FND[;1+NNX-IND[U]])×TRM=0
[26]      TST←⍉(NM,NX)ρTRN
[27]      H←,TSP[((∨⌿3=RXX+TST)=1)/ιNM]
[28]      SUT←(SUT∈H)/SUT
[29]      →(1=MEZ←2*+/EFF←TRM=0)/A9
[30]      CMB←(AMB<MEZ)/AMB
[31]      BINZ←1+EFF
[32]      S←1
[33]   A7:DIF←BINZ⊤CMB[S]
[34]      TRY←TRM+DIF×2-NIM
[35]      TSF←⍉(NZ,NX)ρTRY
[36]      →(0=ρSUF←FSP[((+⌿3=RZZ+TSF)=0)/ιNZ])/S1
[37]      →(MEZ>S←S+1)/A7
[38]   S1:→(DIS=1)/A5
[39]      MAX←+/TRY>0
[40]      H←⊖2-(NXρ2)⊤(SUT-1)
[41]      MEZ←2*+/EFF←(TRM=0)∧DIS∨,∧/H=1⌽H
[42]      →(1=MEZ)/A9
[43]      CMB←(AMB<MEZ)/AMB
[44]      BINZ←1+EFF
```

```
[45]    S←1
[46] S2:DIF←BINZⳁCMB[S]
[47]    TRY←TRM+DIF×2-NIM
[48]    TST←⍉(NZ,NX)ρTRY
[49]    →(0=ρSUF←FSP[((+/3=RZZ+TST)=0)/ιNZ])/S3
[50]    →(MEZ<S←S+1)/S2  ⍪   → A9
[51] S3:→(MAX=+/TRY>0)/A5
[52] A9:→(0<U←U-1)/A2
[53] A6:→(0<NW←NW-1)/A3
[54]    'CYCLE DISSOLUTION'
[55]    DIS←1
[56]    →A4
[57] A5:→(DIS=0)/A1
[58]    TST←⍉(NM,NX)ρTRY
[59]    SUT←,TSP[((+/3=RXX+TST)=0)/ιNM]
[60] A1:G[SUT]←LIM
[61]    CRI←CRI,IND[U]-1
[62]    TST←⍉(NM,NX)ρTRY
[63]    R←R,TRY
[64]    DIS←0
[65]    →A4
[66] A8:S←(ρ(,R))÷NX
[67]    TRI←R←(S,NX)ρR
[68]    NX FORM R
[69]    'CRITICAL SET: ';CRI
     ∇


     ∇ R←X NEWORDER F;U;B;NX;NNX;BINX
[1]     →(NX≠⌊NX←2⊛NNX←ρF)/0
[2]     →(NX≠ρX)/0
[3]     BINX←NXρ2
[4]     R←F
[5]     U←0
[6]  N2:B←⌽BINXⳁU
[7]     R[1+2⊥⌽B[X]]←F[U+1]
[8]     →(NNX>U←U+1)/N2
[9]     →0
     ∇
```

```
      ∇ PRIMIMPLICANT OF;A;U;V;S;RR;IMP;DIS;RS;TRTX;PP;NX
[1]    NX←2⊛ρOF
[2]    'ALGEBRAIC FORM OF THE COMPLEMENTARY FUNCTION:'
[3]    MINIMA(OF=0)
[4]    RR←TRI
[5]    S←(ρRR)[1]
[6]    TRTX←NXρ3
[7]    IMP←¯1+ι3∗NX
[8]    PP←ι0
[9]  P3:DIS←0,ι0
[10]   RS←,RR[S;]
[11]   U←1
[12] P2:→(RS[U]=0)/P1
[13]   DIS←DIS,(DIS+RS[U]×3∗U-1)
[14] P6:→(NX≥U←U+1)/P2
[15]   IMP←(~(IMP∊DIS))/IMP
[16]   →(0<S←S-1)/P3
[17]   →P7
[18] P1:DIS←DIS,(DIS+3∗U-1),(DIS+2×3∗U-1)
[19]   →P6
[20] P7:→(0=ρIMP)/P8
[21]   PP←PP,A←⌊/IMP
[22]   RS←TRTX⊤A
[23]   DIS←A,ι0
[24]   V←NX
[25] P5:→(RS[V]>0)/P4
[26]   DIS←DIS,(DIS+3∗NX-V),(DIS+2×3∗NX-V)
[27] P4:→(0<V←V-1)/P5
[28]   IMP←(~(IMP∊DIS))/IMP
[29]   →P7
[30] P8:PP←⌽⍉TRTX⊤PP
[31]   'SUM OF ALL PRIME IMPLICANTS:'
[32]   NX FORM PP
      ∇


      ∇ SOLVE;MTX;SET;DON;NIC;VEC;LIM;T;FIX;RES;V;MMR
[1]    MTX←⍉(NNX,NNY)ριNNY
[2]    FX←(NY,NNX)ρ0
[3]    DSCR←(NNY,NNX)ρF
[4]    SET←+/DSCR
[5]    DON←(NNY=SET)/¯1+ιNNX
[6]    →(0=ρDON)/S5
[7]    'EACH SOLUTION HAS THE FOLLOWING DONT CARES: ';DON
[8]  S5:NIC←(0=SET)/¯1+ιNNX
[9]    →(0=ρNIC)/S6
```

THE APL PROGRAM "SYSTEM"


```
[10]    'THE NUMBER OF SOLUTIONS IS ZERO UNLESS THE
        FOLLOWING INPUT CONFIGURATIONS ARE FORBIDDEN: ';
        NIC
[11] S6:DIS←(NNY,NNX)ρ((SET≠NNY)∧(SET≠0))
[12]    DIS←DSCR×DIS
[13]    DIS[1;]←(SET=0)∨(SET=NNY)∨DIS[1;]
[14]    SET←+/DIS
[15]    SOL←×/SET
[16]    MTX←DIS×MTX
[17]    'NUMBER OF SOLUTIONS AFTER CONSTRAINTS: SOL= ';SOL
[18]    →(SOL=0)/0
[19]    'DESIRED SOLUTION VECTOR:'
[20]    VEC←,□
[21]    LIM←ρVEC
[22]    T←1
[23] S4:FIX←1+SETτ(VEC[T]-1)
[24]    RES←ι0
[25]    V←1
[26] S1:COL←(MTX[;V]>0)/MTX[;V]
[27]    RES←RES,COL[FIX[V]]
[28]    →(NNX≥V←V+1)/S1
[29]    MMR←(NY,NNX)ρ0
[30]    RES←RES-1
[31]    V←1
[32] S2:MMR[;V]←V×(NYρ2)τRES[V]
[33]    →(NNX≥V←V+1)/S2
[34]    MMR←⊖MMR
[35]    'SOLUTION NUMBER: ';VEC[T]
[36]    V←1
[37] S3:'(Z ';V;') → ';FX[V;]←(0<MMR[V;])+2×(¯1+ιNNX)∈
        (NIC,DON);'           [';¯1+((0<MMR[V;])/MMR[V;]);
        '] ∪ (';DON,NIC;')'
[38]    →(NY≥V←V+1)/S3
[39]    →(LIM≥T←T+1)/S4
     ∇



     ∇ SPACE
[1]     CM←1
[2]     'NUMBER OF Y-VARIABLES:'
[3]     NNY←2*NY←□
[4]     'SYMBOLS FOR Y-VARIABLES: (Y K), (Y̲ K); K= ';ιNY
[5]     'NY= ';NY
[6]     'XY-SPACE SYMBOL:  F[Y;X]'
[7]     XX←⊖(BINX,NYρ2)τ(¯1+ιNNF←2*NX+NY)
[8]     F←NNFρ1
[9]     EQA←1
[10]    'CALL: EQUATION'
     ∇
```

```
      ∇ TABLE
[1]    ⍝ TRUTH TABLE FORMATTING
[2]    EQA←0
[3]    ' NUMBER OF FUNCTIONS:'
[4]    NF←NY←□
[5]    'TABLE IS READY FOR FUNCTIONS  (Z K)  WITH K =';⍳NF
[6]    BINF←NFρ2
[7]    FX←(NF,NNX)ρ0
[8]    TR←TS←0
[9]    F←F←⍳0
[10]   'CALL OFFERINGS: FTRUE, FFALSE, FLIST'
[11]   TEX1←'(ANY LOGICAL APL-MEANINGFUL RELATION WITH
       VARIABLES  (X J), (X J))'
      ∇


      ∇ R←X J
[1]    R←XX[J;]
      ∇


      ∇ R←Y K
[1]    R←XX[NX+K;]
      ∇


      ∇ R←Z L
[1]    R←FX[L;]
      ∇


      ∇ R←X J
[1]    R←~XX[J;]
      ∇


      ∇ R←Y K
[1]    R←~XX[NX+K;]
      ∇


      ∇ R←Z L
[1]    R←(0=,FX[L;])+2×(2=,FX[L;])
      ∇
```

THE APL PROGRAM "SYSTEM"

```
      ∇ ∆FFALSE
[1]     'SUFFICIENT CONDITION OF  F  MUST BE WRITTEN IN
        THE FORM:'
[2]     'F←(LOGICAL RELATION IN  X-VARIABLES)'
[3]     'AND THE EXECUTION OF THIS INSTRUCTION MUST BE
        FOLLOWED'
[4]     'EITHER BY:  FSTOR K'
[5]     'OR BY:  FFALSE:'
      ∇


      ∇ ∆FTRUE
[1]     'SUFFICIENT CONDITION OF  F  MUST BE WRITTEN IN THE'
[2]     'PRESCRIBED FORM:'
[3]     'F←(LOGICAL RELATION IN  X-VARIABLES)'
[4]     'AND THE EXECUTION OF THIS INSTRUCTION MUST BE
        FOLLOWED'
[5]     'BY ONE OF THE PRESCRIBED CALLS:'
[6]     'EITHER:  FSTOR K   OR:  FFALSE'
      ∇


      ∇ ∆SYSTEM
[1]     'SOLVES FOLLOWING PROBLEMS:'
[2]     'TRUTH TABLE GENERATION OR EXISTENCE FUNCTION OF
        THE SYSTEM '
[3]     'TO GENERATE THE TABLE START WITH  LOGIC  AND
        CONTINUE'
[4]     'WITH   TABLE.'
[5]     'TO GENERATE EXISTENCE FUNCTION START WITH  LOGIC'
[6]     'AND THEN CALL  SPACE.'
      ∇


      ∇ ∆TABLE
[1]     'GENERATES TRUTH TABLE IN TWO WAYS:'
[2]     'CALL:  FTRUE OR FFALSE'
[3]     'TO DEFINE BOOLEAN FUNCTIONS BY ALGEBRAIC MEANS.'
[4]     'CALL:  FLIST'
[5]     'TO DEFINE FUNCTIONS BY DECIMAL EQUIVALENTS:'
      ∇
```

Chapter 4
Minimization and Optimization

4.1   INTRODUCTION

To perform minimization and optimization, the
program module sequence of Fig. 4.1 applies.   To begin,
we type:

>)*LOAD* <library number> *OPTIMA*
> *DESIGN*

The interaction is demonstrated in the following
examples.

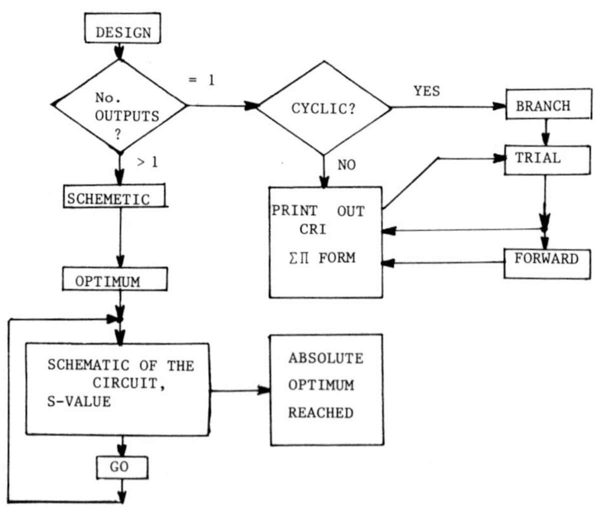

Fig. 4.1. Program module sequence
for minimization and optimization.

4.2.   EXAMPLES

EXAMPLE 4.2.1.   Select a Boolean function of 5 variables at random and minimize its $\Sigma\Pi$-form.   To develop such a function, we call:

```
     EXAMPLE
TYPE NUMBER OF VARIABLES:
□:
     5
MINS→→→→0   2   5   7   12   13   16   19   20   22   26   27   29   30
DONTS→→→3   8   11   15   23   25
```

The function is specified by decimal equivalents. APL vectors MINS and DONTS are mutually exclusive, and they can be used in the program that follows. (MINS represent points where the function is specified as true.) We are ready to call:

```
     DESIGN
NUMBER OF INDEPENDENT VARIABLES OF GIVEN BOOLEAN FUNCTIONS:
□:
     5
THEIR SYMBOLS: ABCDE
SYMBOLISM EXPLANATION:
OK    MEANS:NO MISTAKE
FLT   MEANS: FAULTY TYPING, REQUEST FOR RETYPING
ADD   MEANS: REQUEST FOR ADDITIONAL DATA INSERTION.
```

This explanatory text is intended for students. It is easy to cancel it in the program. In the examples that follow, we will not repeat the printout of explanatory comments. Note that the number of outputs is equal to 1 when we minimize the $\Sigma\Pi$-form of a given function.

```
TYPE THE NUMBER OF OUTPUTS.(BY TYPING: 1 THE PROCEDURE
IS REDUCED TO AN  N-MINIMIZATION OF A SINGLE GIVEN
FUNCTION.)
□:
     1
TYPE-IN DECIMAL EQUIVALENTS OF TRUE MINTERMS
OR THE SYMBOL OF THE CORRESPONDING VECTOR.
□:
     MINS
TYPE: OK OR FLT OR ADD:
□:
     OK
```

TYPE IN UNSPECIFIED MINTERMS (IF NONE, TYPE: NONE).
□:
     DONTS
TYPE: OK OR FLT OR ADD:
□:
     OK
WEIGHT TABLE:

W= 2   FOR MINS: 2   5   12   16   20   26   29   30
W= 3   FOR MINS: 0   19   22
W= 4   FOR MINS: 7   13   27

MARQUAND CHARTS WILL BE USED.

*   CONSULT THE WEIGHT TABLE TO ESTIMATE THE DIFFICULTY OF THE PROBLEM. FOR A FUNCTION OF MORE THAN 6 VARIABLES WHICH HAPPENS TO HAVE ONLY A SMALL COUNT OF LOW WEIGHT MINTERMS (W=0,1,2,3) IT IS ADVISABLE TO BEGIN WITH THE LOWEST WEIGHT VALUE FOUND IN THE TABLE AS A LIMIT, SET FOR THE EXECUTION. THIS LIMIT CAN BE INCREASED LATER ONE BY ONE WHEN IT BECOMES CLEAR THAT THE GENERATION OF NEW TERMS IS ADEQUATE AND THE CORRESPONDING EXECUTION TIME IS ACCEPTABLE.*
STATE OF THE CRITICAL SET:

SET A LIMIT FOR W OR TERMINATE BY TYPING: 0
□:

The weight W has a very simple meaning. It is equal to the <u>count</u> of those nonzeros of the given function which are at a UNIT LOGICAL DISTANCE from a minterm implicant. For instance, $W = 2$ for MIN identified with 5 means: There are exactly two nonzeros of the function at unit logical distance from the MINterm $5 = (00101)_2 \equiv$ <u>EDBC</u>A: $7 = (00111)_2$ and $13 = (01101)_2$ (both belonging to MINS). $W = 4$ for the MINterm $7 = (00111)_2 \equiv$ <u>EDBCA</u> means that there are four nonzeros: $5 = (00101)_2$, a MIN, and $3 = (00011)_2$, $15 = (01111)_2$, and $23 = (10111)_2$, all members of DONTS.

For this example, the limit for W is to be set to 4, so we type:

   4

MINIMIZATION AND OPTIMIZATION

PRESENT STATE:
```
1  0  1  *  0  1  0  1
*  0  0  *  1  1  0  *
1  0  0  1  1  0  1  *
0  *  1  1  0  1  1  0
```

N-MINIMAL FORM:

A CDE   2

A C E   5

AB DE   12

AB DE   16

ABCD   29

ABC   19

AB DE   26

ABC E   22

The N-minimal $\Sigma\Pi$-form of the given incompletely-specified function is:

$$\underline{ACDE} + AC\underline{E} + \underline{ABDE} + \underline{ABDE} + A\underline{B}CD + AB\underline{C} + \underline{AB}DE + \underline{ABCE}$$

$$\{\ 2\ ,\ 5\ ,\ 12\ ,\ 16\ ,\ 29\ ,\ 19\ ,\ 26\ ,\ 22\ \}$$

The corresponding CRITICAL SET is presented below the minimal form. Any pair of MINS taken from this set has components that are MUTUALLY TERM EXLCLUSIVE (MTE). That means that no term implicant T => f of the given function exists that covers both components of any pair. (For functions with a cycle, i.e., CYCLIC functions, the printed set is not always "critical"!) The element of the critical set printed (here) below a term is covered by that term <u>exclusively</u>. For instance, MINterm 19 is covered by AB<u>C</u> and not by any other term of the $\Sigma\Pi$-form.

EXERCISE: Sketch the two-level NAND network defined by the $\Sigma\Pi$-form.

EXAMPLE 4.2.2. Design a two-level NOR network generating the function defined in Example 4.2.1.

Procedure: Develop an N-minimal $\Sigma\Pi$-form for the complement $\bar{f}$ of that function and apply DeMorgan's rules to obtain the minimal $\Pi\Sigma$-form of f. Note that DONTs remain unchanged and that $\bar{f}$ must be true where f was specified as false:

    MINS1←1 4 6 9 10 14 17 18 21 24 28 31

We again call DESIGN (the explanatory texts are not reproduced here):

```
     DESIGN
NUMBER OF INDEPENDENT VARIABLES OF GIVEN BOOLEAN FUNCTIONS:
□:
     5
THEIR SYMBOLS: ABCDE

TYPE THE NUMBER OF OUTPUTS.(BY TYPING: 1 THE PROCEDURE
IS REDUCED TO AN  N-MINIMIZATION OF A SINGLE GIVEN
FUNCTION.)
□:
     1
TYPE-IN DECIMAL EQUIVALENTS OF TRUE MINTERMS
OR THE SYMBOL OF THE CORRESPONDING VECTOR.
□:
     MINS1
TYPE: OK OR FLT OR ADD:
□:
     OK
TYPE IN UNSPECIFIED MINTERMS (IF NONE, TYPE: NONE).
□:
     DONTS
TYPE: OK OR FLT OR ADD:
□:
     OK
WEIGHT TABLE:

W= 0    FOR MINS:   18
W= 1    FOR MINS:   4   28
W= 2    FOR MINS:   6   21   31
W= 3    FOR MINS:   1   10   14   17   24
W= 4    FOR MINS:   9
```

MINIMIZATION AND OPTIMIZATION

STATE OF THE CRITICAL SET:

SET A LIMIT FOR W OR TERMINATE BY TYPING: 0
□:
           4
PRESENT STATE:
0 1 0 ∗ 1 0 1 0
∗ 1 1 ∗ 0 0 1 ∗
0 1 1 0 0 1 0 ∗
1 ∗ 0 0 1 0 0 1

N-MINIMAL FORM:
AB̲CD̲E̲   18

A̲ CD̲E̲   4

AB̲ DE   28

AB̲ D̲E̲   21

ABCD   31

AB̲C̲   1

* B DE̲   10*

    The N-minimal $\Sigma\Pi$-form of $\bar{f}$ is

$$\underline{A}BC\underline{D}\underline{E} + \underline{A}C\underline{D}\underline{E} + \underline{A}B\underline{D}E + \underline{A}\underline{B}\underline{D}\underline{E} + ABCD + \underline{A}\underline{B}\underline{C} + BD\underline{E}$$

$$\{\ 18\ ,\ 4\ ,\ 28\ ,\ 21\ ,\ 31\ ,\ 1\ ,\ 10\ \}$$

and the $\Pi\Sigma$-form of f (by DeMorgan's rules) is

$$f = (A+\underline{B}+C+D+\underline{E})(A+\underline{C}+D+E)(A+B+\underline{D}+\underline{E})$$
$$\cdot\ (\underline{A}+B+D+\underline{E})(\underline{A}+\underline{B}+\underline{C}+\underline{D})(\underline{A}+B+C)(\underline{B}+\underline{D}+E).$$

    EXERCISE: Sketch the corresponding two-level NOR circuit. Discuss which of the two circuits (two-level NAND, two-level NOR) is more practical.

    EXAMPLE 4.2.3.   Design a full adder as a combinational circuit with multiple outputs (see Fig. 1.3). The output functions are: $(Z\ 1) \equiv \{1,2,4,7\}$; $(Z\ 2) \equiv \{3,5,6,7\}$.

The procedure again starts with:

DESIGN
NUMBER OF INDEPENDENT VARIABLES OF GIVEN BOOLEAN FUNCTIONS:
$\square$:
  3
THEIR SYMBOLS: ABC

TYPE THE NUMBER OF OUTPUTS.(BY TYPING: 1 THE PROCEDURE
IS REDUCED TO AN N-MINIMIZATION OF A SINGLE GIVEN
FUNCTION.)
$\square$:
  2
DEFINE FUNCTIONS BY DECIMAL EQUIVALENTS. TYPE: NONE FOR
EMPTY SETS.
FUNCTION LABELED 1 IS TRUE AT:
$\square$:
  1 2 4 7
TYPE: OK OR FLT OR ADD:
$\square$:
  OK
FUNCTION NUMBER 1 IS UNSPECIFIED AT:
$\square$:
  NONE
TYPE: OK OR FLT OR ADD:
$\square$:
  OK
FUNCTION LABELED 2 IS TRUE AT:
$\square$:
  3 5 6 7
TYPE: OK OR FLT OR ADD:
$\square$:
  OK
FUNCTION NUMBER 2 IS UNSPECIFIED AT:
$\square$:
  NONE
TYPE: OK OR FLT OR ADD:
$\square$:
  OK
WEIGHT TABLE:

W= 1 FOR MINS: 9 10 12
W= 2 FOR MINS: 15 19 21 22
W= 5 FOR MINS: 23

STATE OF THE CRITICAL SET:

SET A LIMIT FOR W OR TERMINATE BY TYPING: 0
$\square$:
  5

PRESENT STATE:

```
* * * * * * * *
0 1 1 0 1 0 0 1
0 0 0 1 0 1 1 1    } mosaic function
0 0 0 0 0 0 0 *
```

N-MINIMAL FORM:

AB̲C̲ E̲   9

A̲B̲C̲ E̲   10

A̲B̲C E̲   12

ABC   15

AB D̲   19

A CD̲   21

* BCD̲*   22

*YOU MAY CALL SCHEMATIC. THE RESULTING GRAPH
DESCRIBES THE CIRCUIT UNDER THE FOLLOWING RULES:
    HORIZONTALLY ALIGNED QUADS (---□---) REPRESENT
THE INPUTS OF AN AND-GATE
    HORIZONTALLY ALIGNED CIRCLES (---○---) REPRESENT
AN INPUT OF AN OR-GATE WHOSE OUTPUT VARIABLE IS DESIGNATED
BY A LITERAL PRINTED ABOVE THE CIRCULAR MARKER ○ .*

The purpose of the procedure up until now was to
develop a critical set of the mosaic function. This
critical set is a very important factor of the procedure
that develops the two-level circuit with the minimum of
gates and within this condition with the minimum of gate
inputs. The development of the circuit from the S-minimal
form is understood after a comparison between the N-minimal form (see above) and the circuit printed by the
terminal:

CHAPTER 4

```
        SCHEMATIC
FIRST APPROXIMATION OF THE OPTIMAL NETWORK:
A     A     B     B     C     C           D           E

- | ---□---□--- | ---□--- | -------○------- | --
-□--- | --- | ---□---□--- | -------○------- | --
-□--- | ---□--- | --- | ---□-------○------- | --
- | ---□--- | ---□--- | ---□-------○-------○--
- | ---□--- | ---□--- | --- | ------- | -------○--
- | ---□--- | --- | --- | ---□------- | -------○--
- | --- | --- | ---□--- | ---□------- | -------○--
```

S-VALUE OF THE CIRCUIT: 12
CALL OPTIMUM.

```
        OPTIMUM
```
S-VALUE HAS BEEN REDUCED TO: 11

A B̲C̲ E	9
A̲B̲C E	10
A̲B C E	12
A B C E̲	15
A B D̲	19
A C̲D̲	21
B C D̲	22

AND THE CIRCUIT TO:

```
A     A     B     B     C     C           D           E

- | ---□--- □--- | ---□--- | -------○------- | --
-□--- | --- | ---□---□--- | -------○------- | --
-□--- | ---□--- | --- | ---□-------○------- | --
- | ---□--- | ---□--- | ---□-------○------- | --
- | ---□--- | ---□--- | --- | ------- | -------○--
- | ---□--- | --- | --- | ---□------- | -------○--
- | --- | --- | ---□--- | ---□------- | -------○--
```

S-VALUE OF THE CIRCUIT: 11
TYPE: GO IF YOU WANT TO CONTINUE. IF NOT, TYPE: STOP.
□:
        GO
ABSOLUTE OPTIMUM REACHED.

MINIMIZATION AND OPTIMIZATION

By calling OPTIMUM, the S-minimization is started.
To prevent undesirable cost of execution time, the circuit
is reprinted each time the S-value drops by one unit.
Calling GO can be continued as long as desirable. When
the terminal prints ABSOLUTE OPTIMUM REACHED, the prov-
ably best solution has been reached. (Exception: When
one or more of the functions has a single literal implicant such as A, $\underline{A}$, B,... , the problem must be solved
twice, once in the normal way described previously and
the second time with A, $\underline{A}$, B,... used as the output OR-
gate input and with DONTs corresponding to A, $\underline{A}$, B,...
inserted into functions with single literal implicants.)

EXAMPLE 4.2.4. Design a full adder with outputs
(Z 1), ($\underline{Z}$ 2) (the second output is complemented!). This
time we reproduce the printout without many comments:

```
      DESIGN
NUMBER OF INDEPENDENT VARIABLES OF GIVEN BOOLEAN FUNCTIONS:
□:
      3
THEIR SYMBOLS: ABC

TYPE THE NUMBER OF OUTPUTS.(BY TYPING: 1 THE PROCEDURE
IS REDUCED TO AN  N-MINIMIZATION OF A SINGLE GIVEN
FUNCTION.)
□:
      2
DEFINE FUNCTIONS BY DECIMAL EQUIVALENTS. TYPE: NONE FOR
EMPTY SETS.
FUNCTION LABELED 1 IS TRUE AT:
□:
      1 2 4 7                         ← Function "D"
TYPE: OK OR FLT OR ADD:
□:
      OK
FUNCTION NUMBER 1 IS UNSPECIFIED AT:
□:
      NONE
TYPE: OK OR FLT OR ADD:
□:
      OK
FUNCTION LABELED 2 IS TRUE AT:
□:
      0 1 2 4                         ← Function "E"
```

```
TYPE: OK OR FLT OR ADD:
□:
        OK
FUNCTION NUMBER 2 IS UNSPECIFIED AT:
□:
        NONE
TYPE: OK OR FLT OR ADD:
□:
        OK
WEIGHT TABLE:

W= 1    FOR MINS:   15
W= 2    FOR MINS:   9   10   12
W= 3    FOR MINS:   17   18   20
W= 4    FOR MINS:   16

STATE OF THE CRITICAL SET:

SET A LIMIT FOR  W  OR TERMINATE BY TYPING:   0
□:
        4
PRESENT STATE:
*  *  *  *  *  *  *  *  ◄——— E D
0  1  1  0  1  0  0  1  ◄——————— E D
1  1  1  0  1  0  0  0  ◄——— E D
0  *  *  0  *  0  0  0  ◄——————— E D
```

Note: The mosaic function composed for this purpose
has: The row ED filled with DON'T CAREs (always), the
row ED filled with the function D, the row ED filled with
the function E, and the row ED filled with DON'T CAREs
wherever the product ED is not zero.

In general, for more outputs (for instance, four:
D, E, F, and G), the functions are filled only into rows
with a single non-complemented variable:

	GFED	GFED	GFED	GFED
contains:	G	F	E	D

Any other row contains DON'T CAREs only. For instance, the row GFED contains a DON'T CARE wherever the
product of functions G and E is not equal to zero.

MINIMIZATION AND OPTIMIZATION

N-MINIMAL FORM:
ABC E   15

ABC   9

ABC   10

ABC   12

AB D   16

YOU MAY CALL SCHEMATIC.   THE RESULTING GRAPH
DESCRIBES THE CIRCUIT UNDER THE FOLLOWING RULES:
*   HORIZONTALLY ALIGNED QUADS (---□---) REPRESENT*
THE INPUTS OF AN AND-GATE
*   HORIZONTALLY ALIGNED CIRCLES (---o---) REPRESENT*
AN INPUT OF AN OR-GATE WHOSE OUTPUT VARIABLE IS DESIGNATED
BY A LITERAL PRINTED ABOVE THE CIRCULAR MARKER o .

*   SCHEMATIC*
FIRST APPROXIMATION OF THE OPTIMAL NETWORK:
A   A   B   B   C   C   D   E

```
-|---□---|---□---|---□-------o-------|--
-|---□---□---|---□---|-------o-------o--
-□---|---|---□---□---|-------o-------o--
-□---|---□---|---|---□-------o-------o--
-□---|---□---|---|---|-------|-------o--
```

S-VALUE OF THE CIRCUIT: 12
CALL OPTIMUM.

*   OPTIMUM*
S-VALUE HAS BEEN REDUCED TO: 11
ABC E   15

ABC   9

ABC   10

ABC E   12

AB D   16

AND THE CIRCUIT TO:

A   A   B   B   C   C   D   E

```
-|---□---|---□---|---□-------o-------|--
-|---□---□---|---□---|-------o-------o--
-□---|---|---□---□---|-------o-------o--
-□---|---□---|---|---□-------o-------|--
-□---|---□---|---|---|-------|-------o--
```

S-VALUE OF THE CIRCUIT: 11
TYPE: GO IF YOU WANT TO CONTINUE. IF NOT, TYPE: STOP.
□:
        GO
ABSOLUTE OPTIMUM REACHED.

The resulting circuit (with $(\underline{Z}\ 2)$ as output) is
simpler than the circuit obtained in Example 4.2.3.   It
is used in integrated circuits.

EXAMPLE 4.2.5.   The Boolean function: MINS = 1,
2, 7; DONTs = 0, 3, 5, 6 is at the same time CYCLIC and
ABNORMAL.   Any critical set contains one element at the
maximum, because the elements of pairs: (1,2), (1,7),
(2,7) are MUTUALLY TERM INCLUSIVE (MTI).   The maximal
number of elements in a critical set is M = 1.   The
minimal number of terms in a $\Sigma\Pi$-form, however, is N = 2,
so N > M, a property called ABNORMAL.   We are interested
in the result of the minimization procedure DESIGN applied
to that function:

        DESIGN
NUMBER OF INDEPENDENT VARIABLES OF GIVEN BOOLEAN FUNCTIONS:
□:
        3
THEIR SYMBOLS: ABC

TYPE THE NUMBER OF OUTPUTS.(BY TYPING: 1 THE PROCEDURE
IS REDUCED TO AN N-MINIMIZATION OF A SINGLE GIVEN
FUNCTION.)
□:
        1
TYPE-IN DECIMAL EQUIVALENTS OF TRUE MINTERMS
OR THE SYMBOL OF THE CORRESPONDING VECTOR.
□:
        1 2 7
TYPE: OK OR FLT OR ADD:
□:
        OK
TYPE IN UNSPECIFIED MINTERMS (IF NONE, TYPE: NONE).
□:
        OK
WEIGHT TABLE:

W= 3   FOR MINS: 1   2   7

MINIMIZATION AND OPTIMIZATION

STATE OF THE CRITICAL SET:

SET A LIMIT FOR W OR TERMINATE BY TYPING: 0
☐:
        3
PRESENT STATE:
* 1 1 *
0 * * 1

CYCLIC.   PARTIAL FORM:
RIGHT-HAND-SIDE BORDER INTEGERS
BELONG TO A CRITICAL SET.
RESIDUAL FUNCTION:
* 1 1 *
0 * * 1

TO CONTINUE CALL BRANCH

We were informed that the case is CYCLIC and that
the minimization process would not be started (PARTIAL
FORM is empty).   The residual function is identical to
the function we started with.   Branching is done in the
usual way:

        BRANCH
BRANCHING POINT CHOSEN IS: 1
TABLE OF BRANCHING TERMS:
A   1

   C̲   2

INTEGERS AT THE RIGHT-HAND-SIDE ARE SO CALLED: ROW NUMBERS.
CALL TRIAL.

        TRIAL
SELECT BRANCHING TERM BY TYPING ITS ROW NUMBER.
☐:
        1
BRANCHING TERM:
A   1

COMPLETE COVERAGE.   RESULTING FORM:
A   1

   B   2

RIGHT-HAND-SIDE BORDER INTEGERS
BELONG TO A CRITICAL SET.
LITERALS TOTAL: 2
MAY CALL TRIAL AGAIN.

Because the case is cyclic, the integers printed on the right-hand-side border do not represent a critical set! The N-minimal form, however, is correct. By calling TRIAL again, we can form all branches and compare them:

TRIAL
SELECT BRANCHING TERM BY TYPING ITS ROW NUMBER.
□:
          2
BRANCHING TERM:
   <u>C</u>   1

COMPLETE COVERAGE. RESULTING FORM:
   <u>C</u>   1

A   7

RIGHT-HAND-SIDE BORDER INTEGERS
BELONG TO A CRITICAL SET.
LITERALS TOTAL: 2
MAY CALL TRIAL AGAIN.

Remarks: The cycle is ODD: A, B, <u>C</u> are three existing "primimplicants" forming that cycle. The function is abnormal (M = 1, N = 2). The N-minimal forms A + B, A + <u>C</u> are correct.

EXAMPLE 4.2.6. The completely specified function: MINS = 0, 1, 3, 4, 6, 7, 9, 10, 11, 12, 13, 14 is highly cyclic but normal. Minimize with DESIGN:

DESIGN
NUMBER OF INDEPENDENT VARIABLES OF GIVEN BOOLEAN FUNCTIONS:
□:
          4
THEIR SYMBOLS: ABCD

TYPE THE NUMBER OF OUTPUTS.(BY TYPING: 1 THE PROCEDURE
IS REDUCED TO AN N-MINIMIZATION OF A SINGLE GIVEN
FUNCTION.)
□:
          1
TYPE-IN DECIMAL EQUIVALENTS OF TRUE MINTERMS
OR THE SYMBOL OF THE CORRESPONDING VECTOR.
□:
          0 1 3 4 6 7 9 10 11 12 13 14

MINIMIZATION AND OPTIMIZATION

TYPE: OK OR FLT OR ADD:
☐:
      OK
TYPE IN UNSPECIFIED MINTERMS (IF NONE, TYPE: NONE).
☐:
      NONE
TYPE: OK OR FLT OR ADD:
☐:
      OK
WEIGHT TABLE:

W= 2   FOR MINS: 0   7   10   13
W= 3   FOR MINS: 1   3   4   6   9   11   12   14

STATE OF THE CRITICAL SET:

SET A LIMIT FOR W OR TERMINATE BY TYPING: 0
☐:
      3
PRESENT STATE:
1 1 0 1
1 0 1 1
0 1 1 1
1 1 1 0

CYCLIC. PARTIAL FORM:
RIGHT-HAND-SIDE BORDER INTEGERS
BELONG TO A CRITICAL SET.
RESIDUAL FUNCTION:
1 1 0 1
1 0 1 1
0 1 1 1
1 1 1 0

TO CONTINUE CALL BRANCH

      BRANCH
BRANCHING POINT CHOSEN IS: 0
TABLE OF BRANCHING TERMS:
AB D   1

 BCD   2

INTEGERS AT THE RIGHT-HAND SIDE ARE SO CALLED: ROW NUMBERS.
CALL TRIAL.

      TRIAL
SELECT BRANCHING TERM BY TYPING ITS ROW NUMBER.
☐:
      1
BRANCHING TERM:
AB D   0

COMPLETE COVERAGE. RESULTING FORM:

```
AB D      0   ⎫
                │
A  C      1     │
                │
  BCD     7     ⎬   not a critical set
                │
AB D     10     │
                │
  BCD    13   ⎭
```

RIGHT-HAND-SIDE BORDER INTEGERS
BELONG TO A CRITICAL SET.
LITERALS TOTAL: 14
MAY CALL TRIAL AGAIN.

*   TRIAL*
SELECT BRANCHING TERM BY TYPING ITS ROW NUMBER.
□:
*   2*
BRANCHING TERM:
* BCD   0*

COMPLETE COVERAGE. RESULTING FORM:

```
  BCD     0   ⎫
                │
A  C      4     │
                │
AB D      7     ⎬   critical set (only by checking!)
                │
  BCD    10     │
                │
AB D     13   ⎭
```

RIGHT-HAND-SIDE BORDER INTEGERS
BELONG TO A CRITICAL SET.
LITERALS TOTAL: 14
MAY CALL TRIAL AGAIN.

Remark: The program OPTIMA is well prepared to handle

Boolean Functions with <u>one</u> cycle only. When faced with

functions possessing multiple cycles use the program

SYSTEM.

Chapter 5
The APL Program "OPTIMA"

This chapter consists of the listings of the APL functions which comprise the program OPTIMA. OPTIMA is used for minimization of a Boolean function to the minimal or the optimum $\Sigma\Pi$ or $\Pi\Sigma$ form. It is used for combinational circuits of single or multiple outputs.

OPTIMA is based upon the previously-published theorems of mutual term exclusivity.

```
     ∇ AMPL PRI;X;LIM;U;TRA;Y;TRM;Z
[1]    RRR←TTT←SSS←ι0
[2]    TTT←TTT,G,AA,PRI
[3]    SSS←SSS,COST PRI
[4]    Z←FUND PRI
[5]    RRR←RRR,(-YY+1),(Z∈DOT)/Z
[6]    YY←YY+1
[7]    →(0=Y←+/X←(PRI=0)∧(FCT<0))/0
[8]    TRA←1+X
[9]    LIM←2*Y
[10]   U←1
[11] A1:TRM←PRI+2×(TRAτU)
[12]   TTT←TTT,G,AA,TRM
[13]   SSS←SSS,COST TRM
[14]   Z←FUND TRM
[15]   RRR←RRR,(-YY+1),(Z∈DOT)/Z
[16]   YY←YY+1
[17]   →(LIM>U←U+1)/A1
     ∇
```

```
     ∇ BRANCH
[1]    ⍝ BRANCHING CONTROL
[2]     'BRANCHING POINT CHOSEN IS: ';PIV←MTR[1]
[3]     FUN←FUND SMB←SBOL PIV
[4]     ZCV←(FLS∊FUN)/FLS
[5]     PRIMS 0
[6]     QRS←PRS
[7]     QJ←PJ
[8]     'TABLE OF BRANCHING TERMS:'
[9]     (⍳QJ) PRINT(,QRS)
[10]    'INTEGERS AT THE RIGHT HAND SIDE ARE SO CALLED:
        ROW NUMBERS.'
[11]    'CALL TRIAL.'
     ∇


     ∇ COMB;BMB;P;U
[1]    ⍝ SEQUENCE  AMB  IS MONOTONOUS IN BIT COUNT
[2]     BMB←AMB←2*(¯1+(⍳N))
[3]     P←1
[4]    H1:CMB←⍳0
[5]     U←P
[6]    H2:CMB←CMB,((2*U)+(((2*U)>BMB)/BMB))
[7]     →(N>U←U+1)/H2
[8]     AMB←AMB,CMB
[9]     BMB←CMB
[10]    →(N>P←P+1)/H1
     ∇


     ∇ R←COST X
[1]     R←+/(((FCT>0)∧(1<X))-((FCT<0)∧(1<X)∧(0=2|X)))
     ∇


     ∇ R←CPU
[1]     R←,'⎕⍙CPU=⎕,Q⎕ 60TH''S⎕LI20' ⍙FMT|⍙21-I21
[2]     ⍙21←I21
     ∇


     ∇ COMMENT
[1]     'THIS PROGRAM IS USED FOR COMPUTER AIDED DESIGN
        OF SINGLE AND MULTIPLE OUTPUT COMBINATIONAL CIRCUITS.
        TO BEGIN CALL: DESIGN .'
[2]     ''
```

THE APL PROGRAM "OPTIMA"

```
[3]      'REFERENCE: SVOBODA A., THE CONCEPT OF TERM
         EXCLUSIVENESS AND ITS EFFECT ON THE THEORY OF
         BOOLEAN FUNCTIONS.  JOURNAL OF THE ASSOCIATION
         FOR COMPUTING MACHINERY, VOL. 22,NO. 3,JULY 1975.
             DE VRIES AND SVOBODA, MULTIPLE OUTPUT
         MINIMIZATION WITH MOSAICS OF BOOLEAN FUNCTIONS,
         IEEE TRANSACTIONS ON COMPUTERS, VOL. C-24,NO.8,
         AUG.75.'
      ∇


      ∇ R←DEFIN
[1]    FLT:NONE←R←ι0
[2]    ADD:R←R,□
[3]     'TYPE: OK OR FLT OR ADD:'
[4]      →□
[5]    OK:→0
      ∇


      ∇ DESIGN;U;HR;VT;SET;RED
[1]   ⍝ FIRST PROCEDURE TO BE CALLED.
[2]     ABE←' AABBCCDDEEFFGGHHJJKKLLMMNN'
[3]     'NUMBER OF INDEPENDENT VARIABLES OF GIVEN BOOLEAN
        FUNCTIONS:'
[4]     HR←2*HOR←N←□
[5]     'THEIR SYMBOLS: ';ABE[1+2×ιHOR]
[6]     'SYMBOLISM EXPLANATION:
        OK    MEANS:NO MISTAKE
        FLT   MEANS: FAULTY TYPING, REQUEST FOR RETYPING
        ADD   MEANS: REQUEST FOR ADDITIONAL DATA INSERTION.'
[7]     ''
[8]     'TYPE THE NUMBER OF OUTPUTS.(BY TYPING: 1 THE
        PROCEDURE IS REDUCED TO AN  N-MINIMIZATION OF A
        SINGLE GIVEN FUNCTION.)'
[9]     →(2=VT←2*VRT←□)/F7
[10]    FCT←(VRTρ¨1),(HORρ1)
[11]    MLT←1
[12]    SET←(VRT,HR)ρ0
[13]    U←1
[14]    'DEFINE FUNCTIONS BY DECIMAL EQUIVALENTS. TYPE:
        NONE FOR EMPTY SETS.'
[15]  F1:'FUNCTION LABELED ';U;' IS TRUE AT:'
[16]    TRU←DEFIN
[17]    SET[U;1+TRU]←2
[18]    'FUNCTION NUMBER ';U;' IS UNSPECIFIED AT:'
[19]    TRU←DEFIN
[20]    SET[U;1+TRU]←1
[21]    →(VRT≥U←U+1)/F1
```

```
[22]    RED←(HRρ1),(0<SET[1;])
[23]    U←1
[24] F2:RUD←(HR×2*U)ρ(0<SET[U+1;])
[25]    RED←RED,(RED∧RUD)
[26]    →(VRT>U←U+1)/F2
[27]    RED←(VT,HR)ρRED
[28]    U←0
[29] F3:RED[1+(2*U);]←SET[(U+1);]
[30]    →(VRT>U←U+1)/F3
[31]    TRU←(,RED=2)/⁻1+ιHR×VT
[32]    FLS←(,RED=0)/⁻1+ιHR×VT
[33]    VV←VRT
[34]    →(0<N←HOR+VRT)/F4
[35] F7:MLT←0
[36]    FCT←Nρ1
[37]    'TYPE-IN DECIMAL EQUIVALENTS OF TRUE MINTERMS
        OR THE SYMBOL OF THE CORRESPONDING VECTOR.'
[38]    TRU←DEFIN
[39]    'TYPE IN UNSPECIFIED MINTERMS (IF NONE, TYPE:
        NONE).'
[40]    FLS←TRU,FLS←DEFIN
[41]    FLS←(~((⁻1+ι2*N)∈FLS))/⁻1+ι2*N
[42]    VV←⌊0.5×N
[43] F4:MINIMUM
     ▽
```

```
     ▽ EXAMPLE;W;ALL;Q;JDI
[1]    'TYPE NUMBER OF VARIABLES: ' ⍨
       0 0 ρ6ɪ1,60ιɸ 60 60 60 60 ⊤ɪ20
[2]    ALL←2*☐
[3]    MINS←DONTS←ιQ←0
[4]    JDI←+/? 2 2
[5]    →2+2×JDI
[6]    MINS←MINS,Q
[7]    →10
[8]    DONTS←DONTS,Q
[9]    →10
[10]   →(ALL>Q←Q+1)/4
[11]   W←MINS
[12]   'MINS→→→→';MINS←DONTS
[13]   'DONTS→→→';DONTS←W
     ▽
```

THE APL PROGRAM "OPTIMA"

```
      ∇ EXTEND
[1]      GU← 5 2
[2]    L4:T←0
[3]    L2:→(0<ρTRU)/L3
[4]      →(0<GOT←1)/LL
[5]    L3:→((ρTRU)≥T←T+1)/L5
[6]      →(0<GOT←2)/LL
[7]    L5:AA←A←TRU[T]
[8]     SMB←SBOL TRU[T]
[9]     WT←+/2>SMB
[10]     →(WT≤CLO)/HOP
[11]     →(0<GOT←3)/LL
[12]   HOP:FUN←FUND SMB
[13]    DCV←(FUN∈TRU)/FUN
[14]    ZCV←(FLS∈FUN)/FLS
[15]    PRIMS 0
[16]    DEL←FULSET
[17]    LIST MLT
[18]    →GU[NEXT]
[19]   LL:→0
      ∇
```

```
      ∇ FORWARD
[1]     MTR←TRU
[2]     MST←MST,SYM,SET
[3]     CRS←CRS,PIV,CRI
[4]     'NEW'
[5]     BRANCH
[6]      →0
      ∇
```

```
      ∇ DEL←FULSET;BNC;B;RED;X;XTR
[1]     XTR←,DCV[1]
[2]     DEL←PJρ0
[3]     ALL←PRS>3
[4]     BNC←((Nρ2)⊤DCV[1])
[5]      →(1=B←ρDCV)/0
[6]     J←2
[7]   G2:RED←(PJ,N)ρ(((Nρ2)⊤DCV[J])≠BNC)
[8]    RED←RED∧ALL
[9]     →(0=X←~∧/∨/RED)/G1
[10]    XTR←XTR,DCV[J]
[11]    DEL←DEL∨(∨/RED)×X
[12]  G1:→(B≥J←J+1)/G2
[13]    DCV←XTR
      ∇
```

CHAPTER 5

```
      ∇ FUN←FUND S;U;BNC
[1]      BNC←2|S
[2]      FUN←1ρ(2⊥BNC)
[3]      U←0
[4]      →(N<U←U+1)/0
[5]      →(1<S[U])/4
[6]      FUN←FUN,(FUN+(1-2×BNC[U])×2*N-U)
[7]      →4
      ∇


      ∇ GRAF;TEX;P;V;U;X;Y
[1]      TEX←(8×VRT+HOR)ρ' '
[2]      TEX[¯2+4×(ι(2×VRT+HOR))]←ABE[1+(ι(2×VRT+HOR))]
[3]      TEX[(¯6+8×HOR)+(8×ιVRT)]←' '
[4]      TEX
[5]      ''
[6]      P←1
[7]   S6:LIN←PSM[P;]
[8]      TEX←ι0
[9]      V←1
[10]  S5:Y←2|X←LIN[V]
[11]     U←1
[12]  S4:→((U≠2)∧(U≠6))/S1
[13]     →(V>HOR)/S7
[14]     →(2>X)/S2
[15]     →(2≠5|U+Y)/S2
[16]     TEX←TEX,'□'
[17]     →S3
[18]  S2:TEX←TEX,'|'
[19]  S3:→(8≥U←U+1)/S4
[20]     →(N≥V←V+1)/S5
[21]     TEX
[22]     →(CTR≥P←P+1)/S6
[23]     →30
[24]  S1:TEX←TEX,'-'
[25]     →S3
[26]  S7:→(U≠6)/S1
[27]     →((X≠0)∧(Y=0))/S2
[28]     TEX←TEX,'○'
[29]     →S3
[30]     SSM←+/+/(1<PSM)×((CTR,N)ρ(⌽FCT))
[31]     ''
[32]     'S-VALUE OF THE CIRCUIT: ';SSM
      ∇


      ∇ R←IMPAS
[1]      S←+/(1<PRS)×((PJ,N)ρFCT)
[2]      R←⌊/S
      ∇
```

THE APL PROGRAM"OPTIMA"

```
      ∇ INDX;G;H;X;CNT;SCL;NMN;SUT
[1]     NMN←ρMNR
[2]     H←LYM+X←1
[3]   N2:CNT←TON[H]
[4]     SCL←‾1+ιCNT
[5]     G←1
[6]   N1:SUT←MNR[X]+ι(MNR[X+1]-MNR[X]+1)
[7]     IND[SUT]←SCL[G]
[8]     IMD[SUT]←H
[9]     →(NMN≤X←X+1)/0
[10]    →(CNT≥G←G+1)/N1
[11]    →(0<H←H-1)/N2
      ∇
```

```
      ∇ LIST HOP;PJ;QRS;S;NEG;POS;RED
[1]     →(0<PJ←+/~DEL)/L1
[2]   P8:→(0<NEXT←1)/0
[3]   L1:PRS←(~DEL)/PRS
[4]     →(0=HOP)/P7
[5]     RED←⍉(Nρ2)⊤DCV
[6]     NEG←~∨/RED
[7]     POS←~∨/~RED
[8]     RED←(PJ,N)ρ(2×(NEG∨POS)∧(FCT<0)∧~2|PRS[1;])
[9]     PRS←PRS+2×RED
[10]  P7:DR←IMPAS
[11]    →((BAR<DR)∧(HOP=0))/P8

[12]    PRS←(DR=S)/PRS
[13]    STOR PRS[1;]
[14]    NEXT←2
      ∇
```

```
      ∇ MARQ;RED
[1]     RED←(2*N+1)ρ'* '
[2]     RED[(2×TRU)+1]←'1'
[3]     RED[(2×FLS)+1]←'0'
[4]     ⎕←((2*VV),(2×2*(N-VV)))ρRED
[5]     ' '
      ∇
```

```
      ∇ MINIMUM;X
[1]     CLO←CTR←0
[2]     PRM←CVR←SET←CRI←ι0
[3]     COMB
[4]     ORDER
[5]     ''
[6]     'MARQUAND CHARTS WILL BE USED.
[7]     ''
```

```
[8]      '    CONSULT THE WEIGHT TABLE TO ESTIMATE THE
      DIFFICULTY OF THE PROBLEM. FOR A FUNCTION OF MORE
      THAN  6  VARIABLES WHICH HAPPENS TO HAVE ONLY A
      SMALL COUNT OF LOW WEIGHT MINTERMS (W=0,1,2,3) IT
      IS ADVISABLE TO BEGIN WITH THE LOWEST WEIGHT VALUE
      FOUND IN THE TABLE AS A LIMIT, SET FOR THE EXECUTION.
      THIS LIMIT CAN BE INCREASED LATER ONE BY ONE WHEN
      IT BECOMES CLEAR THAT THE GENERATION OF NEW TERMS
      IS ADEQUATE AND THE CORRESPONDING EXECUTION TIME
      IS ACCEPTABLE.'
[9]      GO← 16 21 10
[10]    'STATE OF THE CRITICAL SET:  ';CRI
[11]    '' × 'SET A LIMIT FOR  W  OR TERMINATE BY TYPING:
      0'
[12]    →(CLO≥X←□)/END × 'PRESENT STATE: ' × MARQ
[13]    CLO←X
[14]    EXTEND
[15]    →GO[GOT]
[16]  WON:'N-MINIMAL FORM:'
[17]    (,CRI) PRINT SET
[18]    →(0=MLT)/0

[19]    'YOU MAY CALL SCHEMATIC. THE RESULTING GRAPH
      DESCRIBES THE CIRCUIT UNDER THE FOLLOWING RULES:
         HORIZONTALLY ALIGNED QUADS (---□---) REPRESENT
      THE INPUTS OF AN AND-GATE
         HORIZONTALLY ALIGNED CIRCLES (---○---) REPRESENT
      AN INPUT OF AN OR-GATE WHOSE OUTPUT VARIABLE IS
      DESIGNATED BY A LITERAL PRINTED ABOVE THE CIRCULAR
      MARKER  ○ .'
[20]    →0
[21]    'CYCLIC. PARTIAL FORM:'
[22]    (,CRI) PRINT SET
[23]    MLUV
[24]    'RESIDUAL FUNCTION:'
[25]    MARQ
[26]    CRS←CRI
[27]    MST←SET
[28]    MTR←TRU
[29]    'TO CONTINUE CALL BRANCH'
[30]    →0
[31]  END:'PARTIAL RESULT:'
[32]    (,CRI) PRINT SET
[33]    MLUV
[34]    MARQ
[35]    →0
      ∇
```

THE APL PROGRAM "OPTIMA"

```
      ∇ MLUV
[1]     'RIGHT HAND SIDE BORDER INTEGERS
        BELONG TO A CRITICAL SET.'
      ∇




      ∇ OPTIMUM;GRP;S;ALL;TER;H;X;MEZ;LIM;TRN;G;SPL;MAX;
        CST;R;GRU;NR;MN;MX;XST;MZ;TRY;LM;TAK;V;MNO;D;TWN;
        E;HOP
[1]     ⍝ RIGOROUS MULTIPLE OUTPUT CIRCUIT DESIGN
        OPTIMIZATION.
[2]     →((⍴DOT)=⍴CRI)/Q9
[3]     GRP←0.5×⍴CVR
[4]     SET←R←ALL←S←⍳0
[5]     YY←X←G←0
[6]   Q1:AA←CVR[1+2×G]
[7]     H←0
[8]     MEZ←CVR[2×1⍳G]
[9]   Q2:TER←PRM[(N×X)+⍳N]
[10]    X←X+1
[11]    AMPL TER
[12]    ALL←ALL,TTT
[13]    S←S,SSS
[14]    R←R,RRR
[15]    →(MEZ>H←H+1)/Q2
[16]    →(GRP>G←G+1)/Q1
[17]    R←R,(-YY+1)
[18]    NR←⍴R
[19]    GRU←⍴S
[20]    TER←(GRU,(N+2))⍴ALL
[21]    LYM←LIM←⌈/IND←TER[;1]
[22]    TRN←(LIM+1)⍴1
[23]    G←0
[24] Q3:TRN[LIM+1-G]←+/IND∊G
[25]    →(LIM≥G←G+1)/Q3
[26]    SPL←IND⍳⌽(¯1+⍳LIM+1)
[27]    MAX←×/TON←TRN
[28]    MNR←(R<0)/⍳NR
[29]    IND←IMD←R
[30]    INDX
[31]    MST←(~(DOT∊CRI))/DOT
[32]    MEZ←⍴MST
[33]    ORG←FEW←FEU←⍳0
[34]    V←1
[35] S0:X←MST[V]
[36]    DEX←(R∊X)/⍳NR
[37]    FEW←FEW,IND[DEX]
[38]    ORG←ORG,((⍴FEW)+1)
[39]    FEU←FEU,IMD[DEX]
[40]    →(MEZ≥V←V+1)/S0
```

```
[41]    MZ←ρORG
[42]    DST←ρFEW
[43]    MAT←(DST,(ρTRN))ρ¯1
[44]    V←1
[45]  S1:MAT[V;FEU[V]]←FEW[V]
[46]    →(DST≥V←V+1)/S1
[47]    V←0
[48]  S6:W←X←1
[49]    VV←TRN⊤V
[50]    →(SSM≤SS←+/S[SPL+VV])/GO
[51]  S3:→(0<+/(MAT[X;]=VV))/S2
[52]    →(ORG[W]>X←X+1)/S3
[53]  GO:→(MAX>V←V+1)/S6
[54]  Q9:'ABSOLUTE OPTIMUM REACHED.'
[55]    →0
[56]  S2:X←ORG[W]
[57]    →(MZ≥W←W+1)/S3
[58]    SSM←SS
[59]    'S-VALUE HAS BEEN REDUCED TO: ';SSM
[60]    TAK←GRUρ0
[61]    TAK[SPL+VV]←1

[62]    SET←TAK≠TER
[63]    CRI←SET[;2]
[64]    SET←(0,0,(Nρ1))/[2] SET
[65]    CRI PRINT(,SET)
[66]    'AND THE CIRCUIT TO:'
[67]    ''
[68]    GRAF
[69]    'TYPE: GO  IF YOU WANT TO CONTINUE. IF NOT, TYPE:
        STOP.'
[70]    →□
[71]  STOP:→0
      ∇
```

```
      ∇ ORDER;T;IND;A;LAB;RED;NTR;U;UU
[1]    ⍝ WEIGHT ORDERING.
[2]    NTR←ρTRU
[3]    LAB←(N+1)ρ0
[4]    RED←NTRρ0
[5]    T←1
[6]  G1:A←TRU[T]
[7]    SMB←SBOL A
[8]    WT←(+/2>SMB)
[9]    IND←+/LAB[ι(WT+1)]
[10]   LAB[WT+1]←LAB[WT+1]+1
[11]   RED[(NTR+1)-(ι(NTR-(IND+1)))]←RED[NTR-(ι(NTR-
       (IND+1)))]
[12]   RED[IND+1]←A
[13]   →(NTR≥T←T+1)/G1
```

THE APL PROGRAM "OPTIMA"

```
[14]    DOT←TRU←RED
[15]    'WEIGHT TABLE:'
[16]    ''
[17]    T←U←1
[18] G3:→(0=UU←LAB[T])/G2
[19]    'W= ';T-1;'    FOR MINS:   ';TRU[(U-1)+ιUU]
[20]    U←U+UU
[21] G2:→((N+1)≥T←T+1)/G3
[22]    ''
     ∇



     ∇ PRIMS HOP;B;C;J;JJ;RED;RAD;ROD;LIM
[1]   ⍝ PRIME IMPLICANTS INCIDENT WITH A POINT.
[2]    TRS←ι0
[3]    →(0<B←ρZCV)/P0
[4]    TRS←TRS,SMB
[5]    →P5
[6] P0:RED←⍉(Nρ2)⊤ZCV

[7]    RAD←(B,N)ρ(2|SMB)
[8]    RAD←RED=RAD
[9]    TRF←1+(SMB<2)
[10]   CMB←((2*WT)>AMB)/AMB
[11] P1:JJ←TRF⊤CMB[1]
[12]   CMB←(~(CMB∈CMB[1]))/CMB
[13]   C←Bρ(+/JJ)
[14]   ROD←(B,N)ρJJ
[15]   ROD←ROD∧RAD
[16]   →(0≠+/(C=+/[2] ROD))/P4
[17]   TRS←TRS,(SMB+(4×JJ))
[18]   →(0<HOP)/P4
[19]   →(0=LIM←ρCMB)/P5
[20]   RED←ι0
[21]   J←1
[22] P3:→(0=+/JJ≠(JJ×(TRF⊤CMB[J])))/P2
[23]   RED←RED,CMB[J]
[24] P2:→(LIM≥J←J+1)/P3
[25]   CMB←RED
[26] P4:→(0<ρCMB)/P1
[27] P5:QJ←PJ←(ρTRS)÷N
[28]   PRS←(PJ,N)ρTRS
[29]   BAR←IMPAS
     ∇
```

```
      ∇ IND PRINT SYM;SEP;NDX;P
[1]     →(0=ρIND)/0
[2]     CTR←(ρSYM)÷N
[3]     NDX←(CTR,N)ρ((2×ιN)-1)
[4]     PSM←φ((CTR,N)ρSYM)
[5]     SEP←1+(PSM>1)×(NDX+(2|PSM))
[6]     P←1
[7]  NXT:ABE[SEP[P;]];'      ';IND[P]
[8]     ''
[9]     →(CTR≥P←P+1)/NXT
      ∇


      ∇ R←SBOL A;U
[1]     R←BNC←(Nρ2)⊤A
[2]     U←1
[3]     →((A+(1-2×BNC[U])×2★N-U)∈FLS)/5
[4]     →6
[5]     R[U]←2+R[U]
[6]     →(N≥U←U+1)/3
      ∇


      ∇ SCHEMATIC
[1]     'FIRST APPROXIMATION OF THE OPTIMAL NETWORK:';''
[2]     GRAF
[3]     'CALL OPTIMUM.'
[4]     →0
      ∇


      ∇ STOR SMB
[1]     SET←SET,SMB
[2]     CRI←CRI,AA
[3]     TRU←TRU[(~(TRU∈DCV))/ιρTRU]
[4]     →(MLT=0)/0
[5]     CVR←CVR,AA,QJ
[6]     PRM←PRM,TRS
      ∇
```

THE APL PROGRAM "OPTIMA"

```
      ∇ TRIAL
[1]    ⍝ BRANCH TRIAL
[2]    'SELECT BRANCHING TERM BY TYPING ITS ROW NUMBER.'
[3]    →(QJ<X←□)/2
[4]    FUN←FUND SYM←QRS[X;]
[5]    DCV←(TRU∈FUN)/TRU←MTR
[6]    'BRANCHING TERM:'
[7]    (,PIV) PRINT SYM
[8]    TRU←TRU[(~(TRU∈DCV))/⍳ρTRU]
[9]    CLO←N
[10]   SET←CRI←⍳0
[11]   EXTEND
[12]   →((GOT=1),(GOT≠1))/ 13 18
[13]   'COMPLETE COVERAGE. RESULTING FORM:'
[14]   (CRS,PIV,CRI) PRINT(MST,SYM,SET)
[15]   MLUV
[16]   'LITERALS TOTAL: ';+/1<MST,SYM,SET
[17]   'MAY CALL TRIAL AGAIN.'
[18]   →0
[19]   'INCOMPLETE COVERAGE. PARTIAL FORM:'
[20]   (CRS,PIV,CRI) PRINT(MST,SYM,SET)
[21]   MLUV
[22]   'LITERALS TOTAL: ';+/1<MST,SYM,SET
[23]   'MAY CALL TRIAL AGAIN TO EXHAUST ALL POSSIBILITIES.
       IF THE FUNCTION IS TOO COMPLEX TO CONTINUE
       EXHUASTIVELY TOWARDS THE MINIMAL COUNT OF LITERALS,
       YOU CAN GIVE UP THE LITERAL MINIMIZATION BY CALLING:
       FORWARD.'
      ∇
```

Chapter 6
Sequential Circuit Aides

6.1 INTRODUCTION

This chapter describes via examples the use of the program BOOL in the study and solution of sequential circuits.

6.2 DESIGNATION OF FUNCTIONS

<u>LIBRARY</u>:)COPY library-number BOOL
<u>BEGIN</u>: BULL

Sequence of functions:

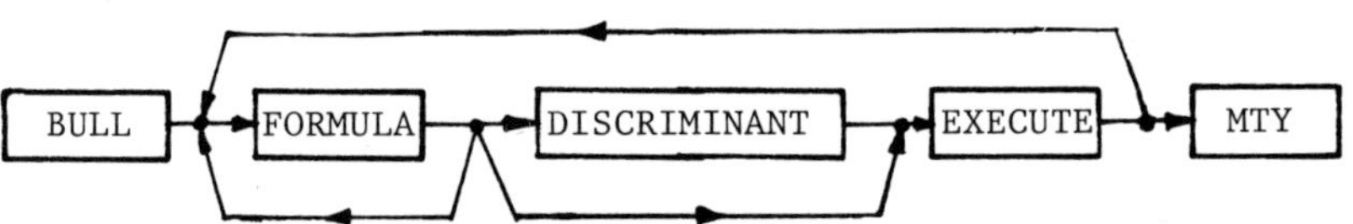

The printout defines the unknowns by their decimal equivalents. In the case where there is no solution, the program prints solutions belonging to restricted input variables (constants). The corresponding configurations of constants' validities become DONT'T CAREs of the solutions. The constant validity configuration for which the unknowns can take on any value is also treated as a DON'T CARE.

For rules concerning the symbolism, see page 12 and Example 8.

The routine EXECUTE modifies the discriminant so that restriction of input variables and other DON'T CAREs

are accomodated.   The   symbol of the modified discriminant is MTY.

6.3   EXAMPLES

EXAMPLE 6.3.1.   A switching network (relay type)
contains a triangle of given switching functions: A,
B, C.   Problem: Find three switching functions D, E, F
that transform the tringle into a star (Fig. 6.1) which
is equivalent to the triangle within the switching network.

The system of equations is solved by calling:

BULL
NUMBER OF CONSTANTS IS:
□:
        3
SYMBOLS FOR CONSTANTS: ABC
NUMBER OF UNKNOWNS IS:
□:
        3
SYMBOLS FOR UNKNOWNS: DEF
CALL FORMULA.

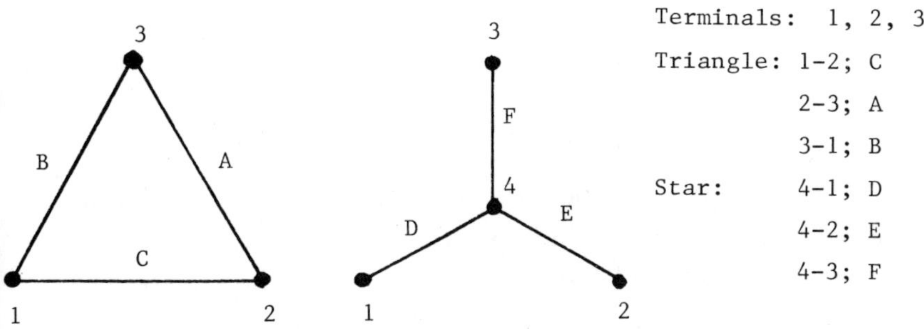

Fig. 6.1.   Switching network
of Example 6.3.1.

```
        FORMULA
WRITE DOWN THE FORMULA:
A+BC=EF
MAY CALL EITHER (NEXT) FORMULA OR DISCRIMINANT OR EXECUTE.

        FORMULA
WRITE DOWN THE FORMULA:
B+CA=FD
MAY CALL EITHER (NEXT) FORMULA OR DISCRIMINANT OR EXECUTE.

        FORMULA
WRITE DOWN THE FORMULA:
C+AB=DE
MAY CALL EITHER (NEXT) FORMULA OR DISCRIMINANT OR EXECUTE.

        DISCRIMINANT
THE DISCRIMINANT VALUE BEFORE CONSTRAINTS IS:
 1 0 0 0 0 0 0 0
 1 0 0 0 0 0 0 0
 1 0 0 0 0 0 0 0
 0 0 0 0 1 0 0 0          ← Number of solutions = 4.
 1 0 0 0 0 0 0 0
 0 0 1 0 0 0 0 0
 0 1 0 0 0 0 0 0
 0 0 0 1 0 1 1 1
AFTER ::EXECUTE:: HAS BEEN CALLED, CALL MTY TO GET
THE CONSTRAINT RECTIFIED DISCRIMINANT.
CALL EXECUTE AFTER ALL CONDITIONS HAVE BEEN PUT IN.
IF NOT, CALL FORMULA AGAIN.
```

The number of solutions of the form

$$D = f\ (A,\ B,\ C)$$
$$E = f\ (A,\ B,\ C)$$
$$F = f\ (A,\ B,\ C)$$

is found by finding the number of nonzeros in each column. The discriminant has four nonzeros in the column belonging to CBA = 1 (at the extreme left) and exactly one nonzero in each of the other columns. We expect four solutions (the product of nonzero counts taken for all columns).

After all of the equations have been written in, we call:

```
        EXECUTE
NUMBER OF SOLUTIONS UNDER ALL CONSTRAINTS:  SOL =  4
```

SEQUENTIAL CIRCUIT AIDES

```
SOLUTION VECTOR:
□:
         1  2  3  4
SOLUTION NUMBER: 1
D = [2   3   4   5   6   7] ∪ ()
E = [1   3   4   5   6   7] ∪ ()
F = [1   2   3   5   6   7] ∪ ()
SOLUTION NUMBER: 2
D = [0   2   3   4   5   6   7] ∪ ()
E = [1   3   4   5   6   7] ∪ ()
F = [1   2   3   5   6   7] ∪ ()
SOLUTION NUMBER: 3
D = [2   3   4   5   6   7] ∪ ()
E = [0   1   3   4   5   6   7] ∪ ()
F = [1   2   3   5   6   7] ∪ ()
SOLUTION NUMBER: 4
D = [2   3   4   5   6   7] ∪ ()
E = [1   3   4   5   6   7] ∪ ()
F = [0   1   2   3   5   6   7] ∪ ()
```

The solutions are printed by listing the ONEs of the functions.

Solution No. 1, the simplest, is very well known. The Marquand charts of that solution are shown in Fig. 6.2.

The other solutions are not as well-known. They are less simple:

No. 2:   D = A + B + C   E = C + A   F = A + B

No. 3:   D = B + C   E = B + C + A   F = A + B

No. 4:   D = B + C   E = C + A   F = C + A + B

EXERCISE: Solve the last example by using the program SYSTEM.

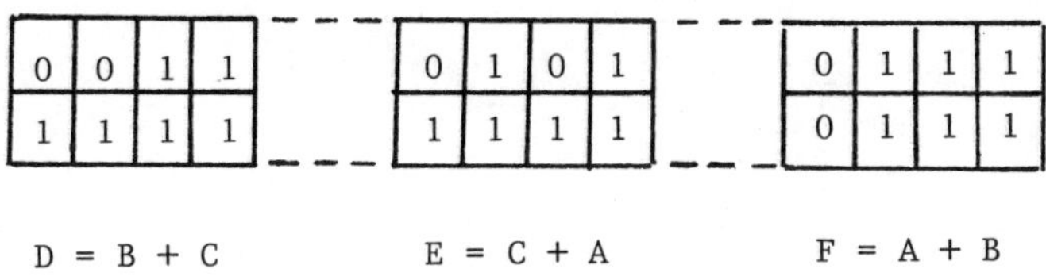

Fig. 6.2. Marquand map of
Example 6.3.1, solution No. 1.

EXAMPLE 6.3.2.   Find all of the steady states, as well as the non-steady-state input configurations, of the sequential circuit shown in Fig. 6.3.

The steady state must satisfy the system of Boolean equations:

$$D = NOR\ (A,E) = \underline{A}\ \underline{E}$$
$$E = AND\ (B,F) = B\ F$$
$$F = OR\ (C,D) = C + D$$

The computer-aided solution using BOOL is found by calling:

```
     BULL
NUMBER OF CONSTANTS IS:
□:
       3
SYMBOLS FOR CONSTANTS: ABC
NUMBER OF UNKNOWNS IS:
□:
       3
SYMBOLS FOR UNKNOWNS: DEF
CALL FORMULA.
```

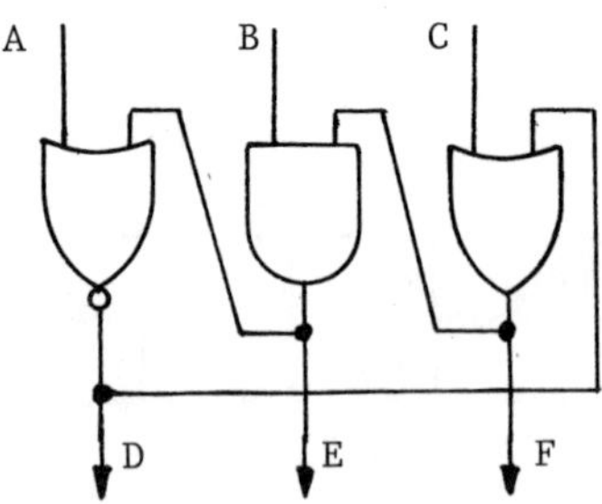

Fig. 6.3.   Sequential
circuit of Example 6.3.2.

FORMULA
WRITE DOWN THE FORMULA:
$D=\underline{A}\underline{E}$
MAY CALL EITHER (NEXT) FORMULA OR DISCRIMINANT OR EXECUTE.

FORMULA
WRITE DOWN THE FORMULA:
E=BF
MAY CALL EITHER (NEXT) FORMULA OR DISCRIMINANT OR EXECUTE.

FORMULA
F=C+D
MAY CALL EITHER (NEXT) FORMULA OR DISCRIMINANT OR EXECUTE.

All equations are in. The existence function of the circuit is obtained by calling:

DISCRIMINANT
THE DISCRIMINANT VALUE BEFORE CONSTRAINTS IS:

0	1	0	1	0	0	0	0
0	0	0	0	0	0	0	0
0	0	0	0	0	0	0	0
0	0	0	0	0	0	0	0
0	0	0	0	0	1	0	0
1	0	0	0	1	0	0	0
0	0	0	0	0	0	1	1
0	0	0	0	0	0	0	0

AFTER ::EXECUTE:: HAS BEEN CALLED, CALL MTY TO GET
THE CONSTRAINT RECTIFIED DISCRIMINANT.
CALL EXECUTE AFTER ALL CONDITIONS HAVE BEEN PUT IN.
IF NOT, CALL FORMULA AGAIN.

Again, the number of solutions of the form

$$D = f(C,B,A)$$
$$E = g(C,B,A)$$
$$F = h(C,B,A)$$

is equal to the product of the number of nonzeros taken for every column:

$$\text{SOL} = 1 \times 1 \times 0 \times 1 \times 1 \times 1 \times 1 \times 1 = 0$$

By constraining the input variables C, B, and A so that the configuration of their validities corresponding to a column filled only by zeros is forbidden, that is $(C\ B\ A)_2 = 2$ (or the $\underline{C}\ B\ \underline{A}$ minterm), exactly one solution is possible. It is interesting that the circuit will behave as a combinational one, as long as the input configuration (C = A = 0, B = 1) is excluded. For this input signal configuration, the circuit will race (oscillate).

In the solution, the forbidden input state is in round parentheses:

```
    EXECUTE
THE NUMBER OF SOLUTIONS IS ZERO, UNLESS
THE FOLLOWING INPUT IDENTIFIERS ARE FORBIDDEN:  2
NUMBER OF SOLUTIONS UNDER ALL CONSTRAINTS:  SOL =  1
SOLUTION VECTOR:
□:
     1
SOLUTION NUMBER: 1
D = [0  4]  ∪ (2)
E = [6  7]  ∪ (2)
F × [0    4    5    6    7]  ∪ (2)
```

The value in parentheses can be proclaimed to be a DON'T CARE because the circuit is not permitted to use it. Then the solution is:

$$D = \underline{A}\,\underline{B}, \qquad E = B\,C, \qquad F = C + \underline{A}$$

The program block BOOL can be used for computer-aided design of sequential circuits. The design procedure mentioned here can be applied to any structural type (model). That means the design of circuits with or without memory elements in the feedback loop, clocked as well as non-clocked circuits, can be assisted by these programs. Roughly speaking, the procedure formulates engineering conditions (constraints) of the circuit in the form of Boolean equations and designs the combinational network by solving that system of equations.

Two examples are offered here. Both illustrate the design of a J-K flip-flop with the trailing edge control. The first example has no memory elements within the feedback loop; the second has memory elements.

EXAMPLE 6.3.3. Design a non-clocked J-K flip-flop with trailing edge control and without memory elements (latches) in the feedback loop.

The type of the circuit is shown in Fig. 6.4. The control (input) variables are A and B. The feedback loop is shown interrupted to permit formulations of constraints in the form of cause-effect relation (implication or equivalence). The split in the feedback permits the use of different variables on each side of the split: C, D; E, F. The variables C, D belong to the inputs of the combinational circuit; E, F belong to its outputs. When the feedbacks are not interrupted, then C must take on the signal level of E (E represents the future signal level of C) and D must take on the signal level of F.

Fig. 6.5 will help us formulate the constraints of the system. The diagram resembles the conventional state diagram and is not a required step in finding a solution.

There are fundamental principles ruling the cause-effect relationship in physics which must be respected to get satisfactory results:

EACH CONSTRAINT MUST BE FORMULATED IN SUCH A WAY THAT THE CAUSE-EFFECT LOGICAL RELATION DEFINES THE TRANSITION OF THE SYSTEM UNIQUELY.

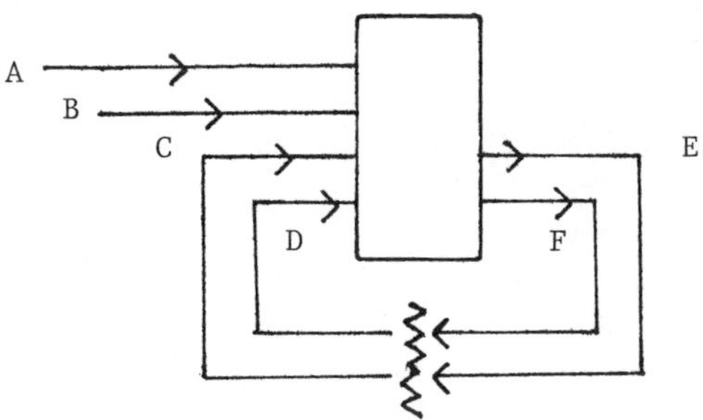

Fig. 6.4. Cause-effect chain.

Explanation: Let us suppose that the system in
Fig. 6.4 is in the state: $S_1 \equiv \{A_1, B_1, C_1, D_1, E_1, F_1\}$
(the feedback is interrupted!). Assuming that the control
A, B is constant and that the structure of the combinational circuit is known, let us inquire about the transition of the circuit if the feedback is not interrupted.
The answer can be formulated by a timeless (purely logical) proposition only if $(E_1 = C_1) \vee (F_1 = D_1)$. For
instance, when $E_1 = C_1$ but $F_1 \neq D_1$, we say simply $D_2 \neq D_1$
(in words: Only D changes the signal level).

Similarly, "only C changes the signal level" is a
proposition that describes completely the transition of
the system when $F_1 = D_1$.

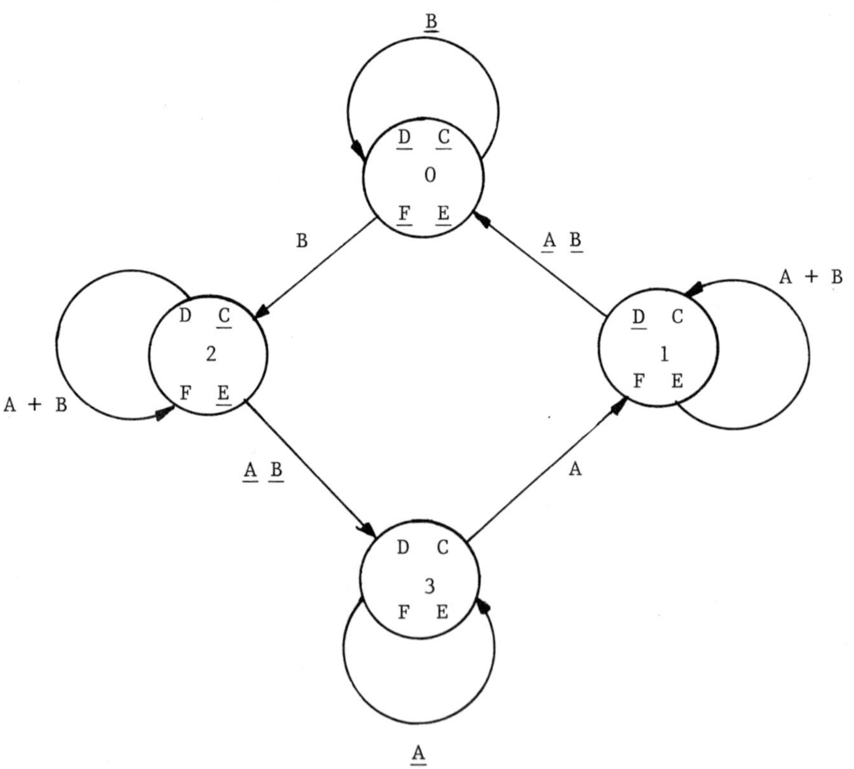

Fig. 6.5. Transition diagram.

If we suppose, however, that $(E_1 \neq C_1)$ AND $(F_1 \neq D_1)$, we are unable to describe the transition by a timeless proposition. Due to physical time delays, the variation in two signal levels will <u>not</u> occur <u>at</u> <u>the</u> <u>same</u> (physical) <u>time</u>.

Definition: The transition of a system is called CONTINUOUS IN LOGICAL TIME if and only if not more than one variable changes during the transition.

By using the continuity notion, the rules for constraint formulation are:

1. The transition induced by a constraint must be continuous in logical time.
2. The control variables may change only when the system is in a steady state.
3. Every transition belonging to a race must be continuous in logical time.

Two feedback loops are suggested in Fig. 6.4. For that reason, four (2^2) possible states of the feedback are at the engineer's disposal. They are represented by circles (Fig. 6.5); identified by state numbers 0, 1, 2, 3; or by state <u>binary</u> identifiers: 00, 01, 10, 11. Binary identifiers indicate the values of the variables of the feedback by the rule

$$i = (D\ C)_2 = (F\ E)_2 \text{ for } i = 0,1,2,3.$$

The principle of continuity in logical time excludes the transitions: $0 \to 3$, $3 \to 0$, $1 \to 2$, and $2 \to 1$, in which two variables change at the same time. For that reason, the circles of the transition diagram are usually sketched in a circular way and numbered by the identifier i in a sequence where only one bit of the binary identifier changes with each step around the circular way. In Fig. 6.5, the sequence in i is 0, 1, 3, 2 (Gray code!).

The transitions are represented in Fig. 6.5 by directional lines (transition arrows) connecting the

circles representing the states of the transition. It
has been said that the binary identifiers of these two
states may disagree in a single bit (or not at all when
the control has no effect on the feedback). The transition arrow can go back to the circle from which it origi-
nated (steady state).

The design procedure is started by choosing a number
n of links in the feedback. That makes 2^n distinct states
of the feedback loop available to the designer. The
designer then tries to establish a transition pattern of
the circuit so that all constraint formulation rules are
respected. Success occurs when there is enough "elbow
room" within the state diagram. If not, a link is added
to the feedback loop of the circuit, and the constraint
formulation is tried again.

In our example, the designer is faced with the
trailing edge control condition. This means that the
output signals, whether true or false, must occur only
after both control signal variables (A, B) have returned
to zero; i.e., when $\underline{A} \cdot \underline{B} = 1$.

The design will be simpler if one of the feedback
links is used to produce the output signal variable. By
choosing the link C-E for that purpose, we observe that
the control signal $\underline{A} \cdot \underline{B}$ must operate with one of the transitions:

1. $(D\ C)_2 = 0\ 1 \leftrightarrow 0\ 0 \equiv 1 \leftrightarrow 0$
2. $(D\ C)_2 = 1\ 1 \leftrightarrow 1\ 0 \equiv 3 \leftrightarrow 2$

The sense of the transition is now chosen. By
choosing $1 \to 0$ or $2 \to 3$, it is clear that the final state
is either 0 or 3 (see Fig. 6.5).

The roles of the control signal variables are now chosen.
If we decide that A = RESET, the output signal is toggled
FALSE in state 0. B $\equiv$ SET results in the output signal
being toggled TRUE in state 3.

Table 6.1. Constraint formulations for Example 6.1.3.

Formulation	Algebraic notation

1. For idle control ($\underline{AB}$ = 1), remain steady at i = 1 ($\underline{DC}$ = 1).
   Constraint: IF ($\underline{AB}$ = 1 and $\underline{DC}$ = 1) THEN ($\underline{EF}$ = 1).
   $\underline{A}\ \underline{B}\ \underline{C}\ \underline{D} \rightarrow \underline{E}\ \underline{F}$

2. For RESET: $\underline{AB}$ = 1, remain steady at i = 0 ($\underline{DC}$ = 1).
   IF ($\underline{AB}$ = 1 and $\underline{DC}$ = 1) THEN ($\underline{EF}$ = 1).
   $\Lambda\ \underline{B}\ \underline{C}\ \underline{D} \rightarrow \underline{E}\ \underline{F}$

3. For control ($\underline{AB}$ = 1), e.g., input B high alone at 0, initiate transition from 0 to 2 by causing the output variables to change so that ($F\underline{E}$ = 1).
   IF ($\underline{AB}$ = 1 and $\underline{DC}$ = 1) THEN ($F\underline{E}$ = 1).
   $\underline{A}\ B\ \underline{C}\ \underline{D} \rightarrow \underline{E}\ F$

4. For a J-K flip-flop, the same transition must be initiated for the control (AB = 1) (both inputs high).
   IF (AB = 1 and $\underline{D}\ \underline{C}$ = 1) THEN ($F\underline{E}$ = 1).
   $A\ B\ \underline{C}\ \underline{D} \rightarrow \underline{E}\ F$

5. After the transition caused by either of the previous constraints, remain steady at i = 2 ($D\underline{C}$ = 1). (Note that $\underline{A}B$ + AB = B.)
   IF ($D\underline{C}$ = 1 and B = 1) THEN ($F\underline{E}$ = 1).
   $B\ \underline{C}\ D \rightarrow \underline{E}\ F$

6. To satisfy the trailing-edge condition, the circuit must wait at i = 2 until both variables A and B return to zero. When B returns to zero sooner than A, the state of the control will be $A\underline{B}$ = 1, and the circuit must remain at i = 2.
   IF ($A\underline{B}$ = 1 and $\underline{CD}$ = 1) THEN ($\underline{EF}$ = 1).
   $A\ \underline{B}\ \underline{C}\ D \rightarrow \underline{E}\ \underline{F}$

Table 6.1 (Continued).

Formulation	Algebraic notation

7. Trailing edge $\underline{A}\underline{B}$ = 1 when the system is at i = 2 causes transition 2 → 3 by making EF = 1.

IF ($\underline{A}\underline{B}$ = 1 and $\underline{C}$D = 1) THEN (EF = 1).

$\underline{A}\ \underline{B}\ \underline{C}\ D \to E\ F$

8. For idle control ($\underline{A}\underline{B}$ = 1), remain steady at i = 3.

IF ($\underline{A}\underline{B}$ = 1 and CD = 1) THEN (EF = 1).

$\underline{A}\ \underline{B}\ C\ D \to E\ F$

9. For SET ($\underline{A}$B = 1), remain steady at i = 3 (CD = 1).

$\underline{A}\ B\ C\ D \to E\ F$

10. For control $A\underline{B}$ = 1 (input A high alone) at i = 3, initiate transition 3 → 1 by causing $E\underline{F}$ = 1.

$A\ \underline{B}\ C\ D \to E\ \underline{F}$

11. Initiate the same transition at i = 3 for the control AB = 1 (both inputs high).

$A\ B\ C\ D \to E\ \underline{F}$

12. After a transition caused by either of the two previous constraints, remain steady at i = 1 ($C\underline{D}$ = 1).

$A\ C\ D \to E\ \underline{F}$

13. To satisfy the trailing-edge condition, wait at i = 1 until $\underline{A}\underline{B}$ = 1. When A returns to zero sooner than B, the state of the control will be $\underline{A}\underline{B}$ = 1 but the circuit must stay at i = 1.

$A\ B\ C\ \underline{D} \to E\ \underline{F}$

14. Trailing edge $\underline{A}\underline{B}$ = 1 when the system is at i = 1 causes transition 1 → 0 by making $\underline{E}\underline{F}$ = 1.

$A\ \underline{B}\ C\ \underline{D} \to \underline{E}\ \underline{F}$

Keeping in mind that, as a J-K flip-flop, the circuit must always go from TRUE to FALSE (or vice-versa) when both inputs are stimulated (for AB = 1), we are ready to formulate the constraints. Refer to Table 6.1.

Implications that have identical expressions on their right sides can be merged into one implication. For instance, the constraints

1, 2, 14 merge into:

$$\underline{A}\ \underline{B}\ \underline{C}\ \underline{D} + A\ \underline{B}\ \underline{C}\ \underline{D} + \underline{A}\ \underline{B}\ C\ \underline{D} \to \underline{E}\ \underline{F}$$
$$\equiv \underline{B}\ \underline{C}\ \underline{D} + \underline{A}\ \underline{B}\ \underline{D} \to \underline{E}\ \underline{F}$$

3, 4, 5, 6 merge into:

$$\underline{A}\ B\ \underline{C}\ \underline{D} + A\ B\ \underline{C}\ \underline{D} + A\ \underline{B}\ \underline{C}\ D + B\ \underline{C}\ D \to \underline{E}\ F$$
$$= A\ \underline{C}\ D + B\ \underline{C} \to \underline{E}\ F$$

7, 8, 9 merge similarly into:

$$\underline{A}\ \underline{B}\ D + \underline{A}\ C\ D \to E\ F$$

10, 11, 12, 13 merge finally into:

$$A\ C + B\ C\ \underline{D} \to \underline{E}\ \underline{F}$$

To design the combinational network, call BULL. Note that the outputs E, F are sought as Boolean functions of the inputs A, B, C, D.

```
     BULL
NUMBER OF CONSTANTS IS:
□:
      4
SYMBOLS FOR CONSTANTS: ABCD
NUMBER OF UNKNOWNS IS:
□:
      2
SYMBOLS FOR UNKNOWNS: EF
CALL FORMULA.
```

The FORMULA call is repeated once for each of the four implications:

```
     FORMULA
WRITE DOWN THE FORMULA:
BCD+ABD→EF
MAY CALL EITHER (NEXT) FORMULA OR DISCRIMINANT OR EXECUTE.
```

```
        FORMULA
WRITE DOWN THE FORMULA:
ADC+BC→FE
MAY CALL EITHER (NEXT) FORMULA OR DISCRIMINANT OR EXECUTE.

        FORMULA
WRITE DOWN THE FORMULA:
DAB+DCA→EF
MAY CALL EITHER (NEXT) FORMULA OR DISCRIMINANT OR EXECUTE.

        FORMULA
WRITE DOWN THE FORMULA.
AC+BCD→EF
MAY CALL EITHER (NEXT) FORMULA OR DISCRIMINANT OR EXECUTE.
```

If the DISCRIMINANT has a column full of zeros, no solution exists unless the set of input signal configurations for that column is forbidden. A column full of ones means "don't care" for the corresponding input signal configuration. In both cases, the discriminant is modified to produce solutions with don't cares for input configurations for columns with either all ones or all zeros (these are listed within the round parentheses of the printout).

```
        DISCRIMINANT
THE DISCRIMINANT VALUE BEFORE CONSTRAINTS IS:
 1 1 0 0 1 0 0 0 0 0 0 0 0 0 0 0
 0 0 0 0 0 1 1 1 0 0 0 0 0 1 0 1
 0 0 1 1 0 0 0 0 0 1 1 1 0 0 0 0
 0 0 0 0 0 0 0 0 1 0 0 0 1 0 1 0
AFTER ::EXECUTE:: HAS BEEN CALLED, CALL MTY TO GET
THE CONSTRAINT RECTIFIED DISCRIMINANT.
CALL EXECUTE AFTER ALL CONDITIONS HAVE BEEN PUT IN.
IF NOT, CALL FORMULA AGAIN.
```

In the present case, the discriminant has none of the singularities previously mentioned. Each column contains exactly one non-zero. Exactly one solution exists for our problem:

```
        EXECUTE
NUMBER OF SOLUTIONS UNDER ALL CONSTRAINTS:  SOL =   1
SOLUTION VECTOR:
□:
        1
SOLUTION NUMBER: 1
E = [5   6   7   8   12   13   14   15] ∪ ()
F = [2   3   8   9   10   11   12   14] ∪ ()
```

SEQUENTIAL CIRCUIT AIDES

The following steps are recommended from this point on:

1. Find N-minimal ΣΠ-forms of both (all) output func-
    tions (here, E and F).
2. Find all prime implicants of those functions.
3. Design the circuit.
4. Draw DYNAMIC SCHEMATIC to look for hazards.
5. Go through exhaustive hazard analysis.
6. Eliminate hazards.

   The program block SYSTEM is used to complete steps
1 and 2.

```
      LOGIC
NUMBER OF X-VARIABLES:
□.
      4
NX= 4
SYMBOLS FOR X-VARIABLES:  (X J),  (X̲ J); J=1   2   3   4
CALL:  TABLE, SPACE
      TABLE
 NUMBER OF FUNCTIONS:
□:
      2
TABLE IS READY FOR FUNCTIONS  (Z K)  WITH K= 1   2
CALL OFFERINGS: FTRUE, FFALSE, FLIST
      FLIST
 DECIMAL EQUIVALENTS OF ONES (AT LEAST ONE ITEM):
□:
      5  6  7  8  12  13  14  15
 DECIMAL EQUIVALENTS OF DONT CARES:
□:
      ι0
 CALL:  FSTOR  K (WHERE K IS WELL SPECIFIED)
      FSTOR 1
 CALL:  FTRUE,  FFALSE,  FLIST  TO DEFINE THE NEXT
      FUNCTION  (F K)  WITH  K=2
      FLIST
 DECIMAL EQUIVALENTS OF ONES (AT LEAST ONE ITEM):
□:
      2  3  8  9  10  11  12  14
 DECIMAL EQUIVALENTS OF DONT CARES:
□:
      ι0
 CALL:  FSTOR  K (WHERE K IS WELL SPECIFIED)
      FSTOR 2
 TRUTH TABLE IS READY:
MAY CALL:  FTRUE, FFALSE,  FLIST FOR  K= 3 ≤2
MAY EXECUTE  FX  TO PRINT THE TABLE
```

The first row of the truth table contains (Z 1), designated by E in Fig. 6.4. The second row contains (Z 2), designated by F.

```
     FX
 0 0 0 0 0 1 1 1 1 0 0 0 1 1 1 1
 0 0 1 1 0 0 0 0 1 1 1 1 1 0 1 0
```

To minimize, we call:

MINIMA (Z 1)
$\underline{A}\underline{B}D + BC + AC$
CRITICAL SET: 8   6   5

MINIMA (Z 2)
$\underline{A}D + \underline{C}D + B\underline{C}$
CRITICAL SET: 14   9   3

PRIMIMPLICANT (Z 1)
ALGEBRAIC FORM OF THE COMPLEMENTARY FUNCTION:
$\underline{A}\underline{B}D + B\underline{C} + A\underline{C}$
CRITICAL SET: 4   10   9
SUM OF ALL PRIME IMPLICANTS:
$A\underline{C} + BC + \underline{A}\underline{B}D + CD$

PRIMIMPLICANT (Z 2)
ALGEBRAIC FORM OF THE COMPLEMENTARY FUNCTION:
$AC + C\underline{D} + \underline{B}\underline{D}$
CRITICAL SET: 15   6   1
SUM OF ALL PRIME IMPLICANTS:
$B\underline{C} + \underline{A}D + \underline{C}D$

MINIMA ($\underline{Z}$ 1)
$\underline{A}\underline{B}D + B\underline{C} + A\underline{C}$
CRITICAL SET: 4   10   9

MINIMA ($\underline{Z}$ 2)
$AC + C\underline{D} + \underline{B}\underline{D}$
CRITICAL SET: 15   6   1

PRIMIMPLICANT ($\underline{Z}$ 1)
ALGEBRAIC FORM OF THE COMPLEMENTARY FUNCTION:
$\underline{A}\underline{B}D + BC + AC$
CRITICAL SET: 8   6   5
SUM OF ALL PRIME IMPLICANTS:
$A\underline{C} + B\underline{C} + \underline{A}\underline{B}D + \underline{C}D$

PRIMIMPLICANT ($\underline{Z}$ 2)
ALGEBRAIC FORM OF THE COMPLEMENTARY FUNCTION:

SEQUENTIAL CIRCUIT AIDES

$\underline{A}D + \underline{C}D + B\underline{C}$
CRITICAL SET: 14   9   3
SUM OF ALL PRIME IMPLICANTS:
AC + $\underline{B}D$ + C$\underline{D}$

The shematic of the sequential circuit (step 4) is
now drawn, preferably in the form called DYNAMIC SCHEMATIC
(Fig. 6.6).   Let us start with the variant implementing
MINIMA (Z 1) and MINIMA (Z 2).   There is one vertical line

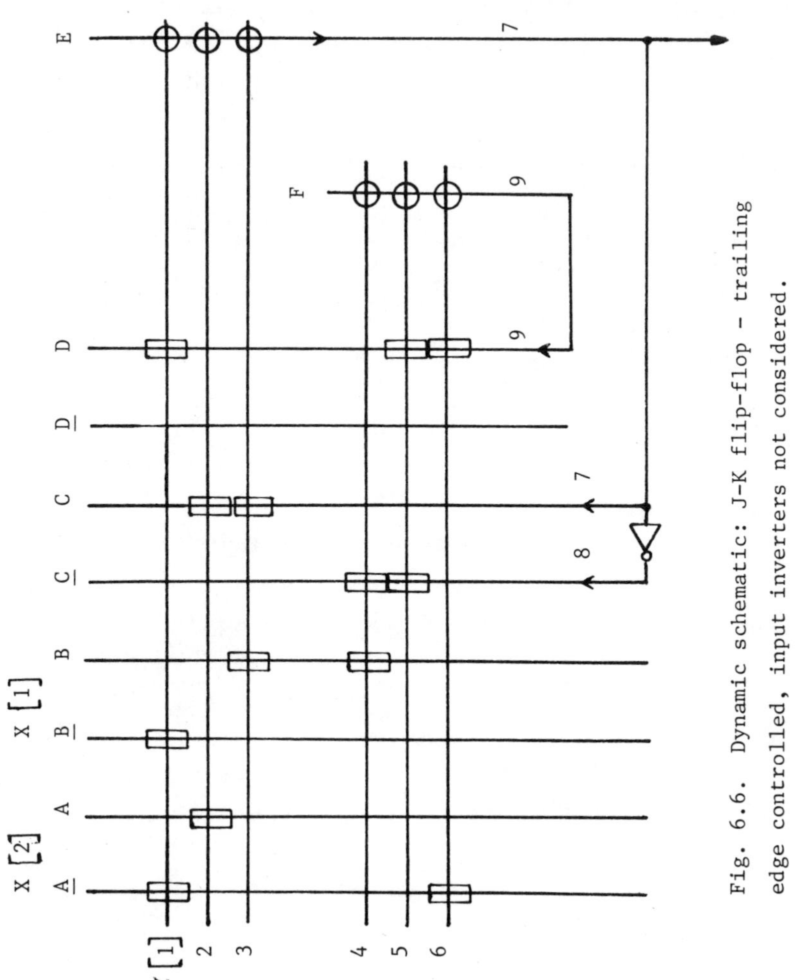

Fig. 6.6. Dynamic schematic: J-K flip-flop – trailing edge controlled, input inverters not considered.

drawn to represent each of the variables A, $\underline{A}$, B, $\underline{B}$, C, $\underline{C}$, D, $\underline{D}$, E, and F. Lines for $\underline{E}$ and $\underline{F}$ are not drawn since the expressions do not include them:

$$E = \underline{AB}D + BC + AC, \quad F = \underline{A}D + \underline{C}D + B\underline{C}.$$

A horizontal line is drawn for each term appearing on the right side of the expressions. In the two-level AND-OR implementation, each line will belong to an AND gate. The input signals entering such an AND gate are represented by markings in the form of small rectangles (for instance, the horizontal line at the top stands for an AND gate with inputs $\underline{A}$, $\underline{B}$, D). Vertical lines E and F, representing the output signals of the combinational network, also represent the corresponding output OR gate. Small circular markings designate which AND gate outputs are used as the OR gate inputs. In Fig. 6.6, the OR gate producing E is entered by the outputs of three AND gates (lines 1, 2, and 3).

In combination with a set of small rods*, a graphic-mechanical working model of the sequential circuit can be created that is well suited for use in analyzing the activity of the circuit. When a rod is placed to cover a vertical line representing a variable (for instance, $\underline{A}$), it means that that particular variable is ON (high). In Fig. 6.7, assume that four rods are placed to cover vertical lines $\underline{A}$, $\underline{B}$, C, and D. Some of the square markers become masked, some remain unmasked. All markings of the horizontal line No. 1 are masked to indicate that all inputs of the AND gate represented by the horizontal line are ON so that the output of the AND gate must be ON. Similarly, the output of the AND gate represented by line

----

* In the classroom, coffee stirrers have been used for this purpose.

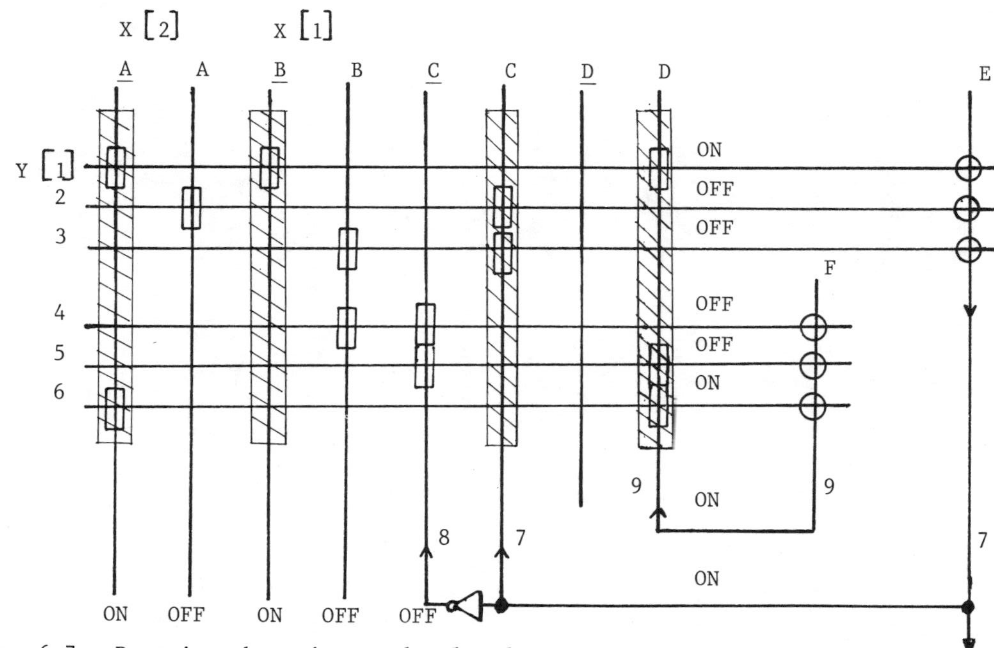

Fig. 6.7. Dynamic schematic: rods placed to represent the steady state $\underline{A}\ \underline{B}\ C\ D\ E\ F = 1$. Note the the variables A, B, C are represented by two wires: A, $\underline{A}$; B, $\underline{B}$; C, $\underline{C}$. The transition from $\underline{A}$ to A in the case shown will cause a hazard if the wire $\underline{A}$ becomes low sooner than A becomes high. In that case E could become illegally low.

No. 6 is ON because all markers of that line are masked. Lines No. 2, 3, 4, and 5 are OFF because at least one square marker is visible in incidence with each of these lines. One of the inputs of the OR gate belonging to the output E being ON (vertical line No. 7), therefore E is ON. If E is ON then C is also ON but line No. 8 is OFF due to the inverter. Similarly, the OR gate belonging to the output F has one input ON so that line No. 9 must be ON. The model of the circuit represents the state of the circuit designated by i = 3 in Fig. 6.5 with idle control ($\underline{AB}$ = 1).

The dynamic schematic can be used in this way to analyze the performance of the circuit perfunctorily or with gradually increasing care about detailed impulse-timing information.

The simplest use of the dynamic schematic is the checking of equations. Transitions between states can be observed without trying to discover hazards. For instance, starting with the state shown in Fig. 6.6, we want to observe the transition due to the change of the control from $\underline{A}$ $\underline{B}$ to A $\underline{B}$. To do it, the rod that masks the vertical line $\underline{A}$ is moved to mask the line A. Now the schematic is inspected: Horizontal line No. 1 now shows a rectangular marker so that the output of the corresponding AND gate must be OFF (it was ON). All markers of horizontal line No. 2 are masked so that the output of the corresponding AND gate must be ON (it was OFF). Horizontal line No. 6 shows a marker, so its AND gate's output is OFF (it was ON). The OR gate whose output is line No. 9 has all inputs OFF. For that reason, line No. 9 is OFF (it was ON). The rod masking line No. 9 must be moved to mask the verical line $\underline{D}$. The last motion causes no change in the ON-OFF signal distribution. The circuit is stable with $C\underline{D}$ = $E\underline{F}$ = 1 after correct transition with i = 3 $\rightarrow$ 1.

SEQUENTIAL CIRCUIT AIDES

The animation of the dynamic schematic can be used
to detect hazards. Let us go through the preceding exp-
eriment more carefully. The motion of the rod from $\underline{A}$ to
A caused a <u>change</u> <u>in</u> <u>two</u> <u>lines</u> <u>at</u> <u>the</u> <u>same</u> <u>time</u>: Line
No. 1 went OFF, and line No. 2 went ON. Those two lines
represent two AND gate outputs; the reaction times (time
delay) of those gates are always different; thus, we must
consider the case where AND gate No. 2 is very much slower
than AND gate No. 1. In that case, line No. 1 will be OFF
long before line No. 2 goes ON. The output of the OR gate
controlling line No. 7 will go OFF. To represent that
fact by animation, the rod masking line C ($\equiv$ No. 7) must
be removed. Another marker of line No. 2 becomes unmasked
so that the line will be prevented from going ON. The
result represents a hazard: The circuit went through i =
$3 \rightarrow 2 \rightarrow 0$ or through i = $3 \rightarrow 1 \rightarrow 0$, depending on the reaction
time of the AND gate represented by line No. 6. The final
state of both transitions is spurious. The designed version of the circuit has a hazard for the transition considered.

Input signal inverters are added in Fig. 6.8. This
schematic should be animated in a way that shows the role
of the delay in the inverters. EXERCISE: Show that the
signal B inverter induces hazards.

Fig. 6.9 illustrates ways to eliminate hazards. The
hazard due to the difference in reaction times between
the AND gates implementing the terms $\underline{A}\ \underline{B}\ D$ and A C of the
expression E = $\underline{A}BD$ + BC + AC is eliminated by adding the
term CD (a prime implicant of E) to that expression so
that E = CD + $\underline{A}BD$ + BC + AC (see the result of PRIMIMPLI-
CANT (Z 1) !), and by adding a corresponding AND gate to
the circuit (line No. 1 in Fig. 6.9). The output of this
gate remains ON during the transition i = $3 \rightarrow 1$ so that
the signal changes at the outputs of the gates corresponding to the terms $\underline{A}BD$ and AC do not influence the output

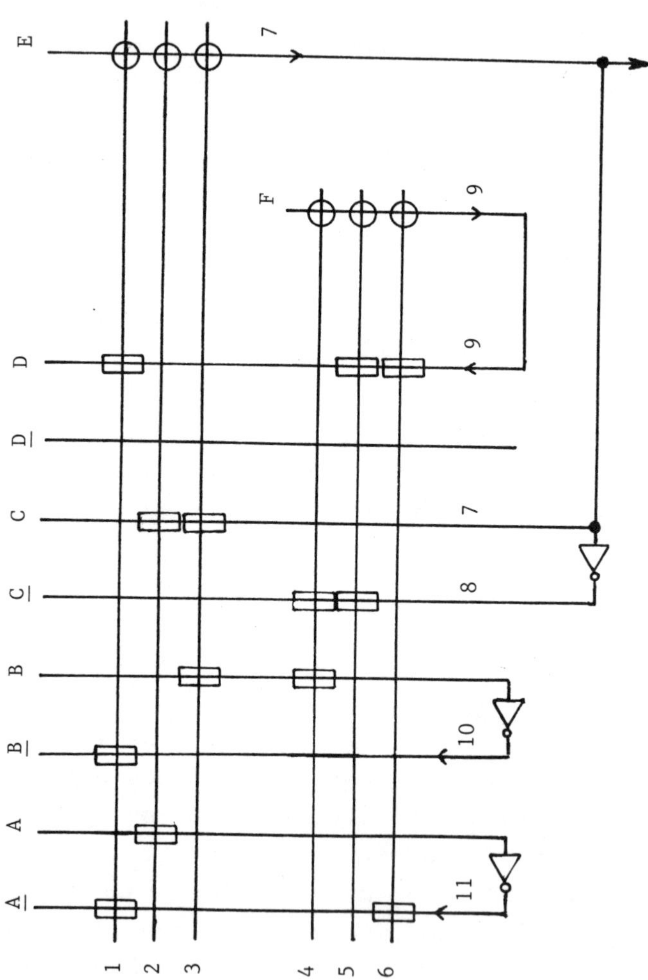

Fig. 6.8. Dynamic schematic: J-K flip-flop – trailing edge controlled, input inverters induce hazards.

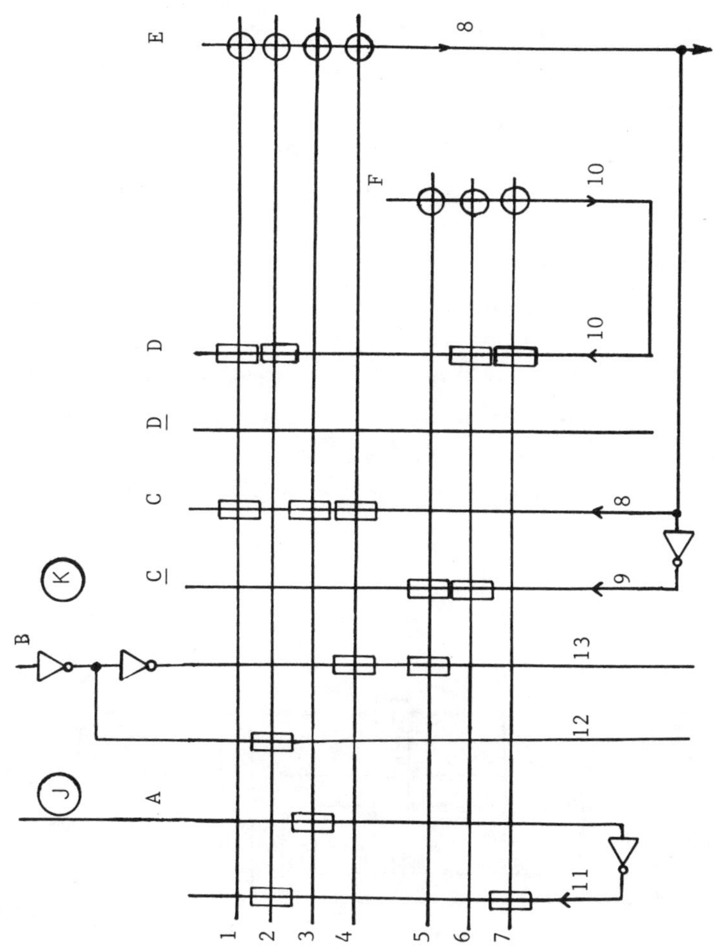

Fig. 6.9. Dynamic schematic: J-K flip-flop – trailing edge controlled, hazards eliminated. Speed independant circuit.

E.   The hazard due to the delay of the B inverter shown in Fig. 6.8 is eliminated by the cascading of two inverters, as shown in Fig. 6.9, which causes line No. 12 to always change its signal level <u>before</u> line No. 13.

EXAMPLE 6.3.4.   Design a nonclocked J-K flip-flop with trailing-edge control but with memory elements (S-R NOR latches) inserted in the feedback loop.   A model with memory elements in the feedback loop is very attractive because it can be proven that there will be no hazards in the designed network if the logical time continuity design principles were respected during its design.

Fig. 6.10 shows the model of the circuit used in this example.   Two S-R flip-flops make the labeling of the variables quite easy: The input signals C and D of the combinational circuit are derived from the outputs of the flip-flops together with their complements, $\underline{C}$ and $\underline{D}$ (output signal inverters are eliminated).   Four control signals, E, F, G, and H, must be provided to control the flip-flops.

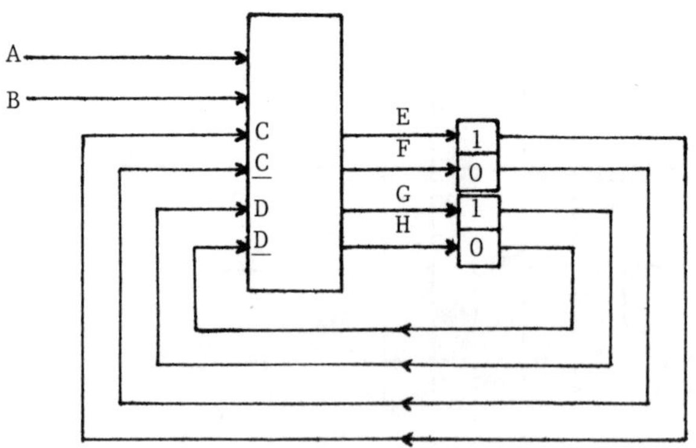

Fig. 6.10.   Sequential circuit with memory elements (S-R flip-flops) in the feedback loop.

Fig. 6.11 shows the state diagram of the sequential circuit. The diagram is almost identical to that of Fig. 6.5; the trigger variables E, F, G, and H are not marked because they are not rigidly related to C and D.

The constraints formulated for the circuit are shown in Table 6.2.

To solve the set of simultaneous implications, we call:

```
     BULL
NUMBER OF CONSTANTS IS:
□:
     4
SYMBOLS FOR CONSTANTS: ABCD
NUMBER OF UNKNOWNS IS:
□:
     4
SYMBOLS FOR UNKNOWNS: EFGH
CALL FORMULA.
     FORMULA
WRITE DOWN THE FORMULA:
BCD→EG
MAY CALL EITHER (NEXT) FORMULA OR DISCRIMINANT OR EXECUTE.
     FORMULA
WRITE DOWN THE FORMULA:
BCD→GHE
MAY CALL EITHER (NEXT) FORMULA OR DISCRIMINANT OR EXECUTE.
     FORMULA
WRITE DOWN THE FORMULA:
ADC+BDC→EH
MAY CALL EITHER (NEXT) FORMULA OR DISCRIMINANT OR EXECUTE.
     FORMULA
WRITE DOWN THE FORMULA:
DABC→EFH
MAY CALL EITHER (NEXT) FORMULA OR DISCRIMINANT OR EXECUTE.
     FORMULA
WRITE DOWN THE FORMULA:
CDA→FH
MAY CALL EITHER (NEXT) FORMULA OR DISCRIMINANT OR EXECUTE.
     FORMULA
WRITE DOWN THE FORMULA:
ACD→HGF
MAY CALL EITHER (NEXT) FORMULA OR DISCRIMINANT OR EXECUTE.
     FORMULA
WRITE DOWN THE FORMULA:
ACD+BCD→FG
MAY CALL EITHER (NEXT) FORMULA OR DISCRIMINANT OR EXECUTE.
```

Table 6.2. Constraint formulations for Example 6.1.2.

Formulation	Algebraic notation
1. The circuit remains steady at $i = 0$ ($\underline{CD}$ = 1) for idle control ($\underline{AB}$ = 1) and for RESET ($A\underline{B}$ = 1). [Note that both flip-flops are reset (output at 0) to produce $\underline{CD}$. Signal E or G would set a flip-flop and change $\underline{CD}$ = 1.]	$\overline{B}\ \overline{C}\ \overline{D} \rightarrow \overline{E}\ \overline{G}$
2. The control ($\underline{AB}$ = 1) or ($A\underline{B}$ = 1) at $i = 0$ ($\underline{CD}$ = 1) must cause the transition $i = 0 \rightarrow 2$. [Note: To make D signal ON, the controlling flip-flop must be switched ON (output at 1): Trigger signal G, no trigger signal H. The other flip-slop must remain reset: No trigger signal E (anything at F).]	$B\ \underline{C}\ \underline{D} \rightarrow G\ \underline{H}\ \underline{E}$
3. The transition $i = 2 \rightarrow 3$ may proceed only when ($A\underline{B}$ = 1) = trailing-edge condition. Thus, the system must remain steady at $i = 2$ ($\underline{C}D$ = 1) as long as ($A+B = 1$).	$(A + B)\ \underline{C}\ D \rightarrow \underline{E}\ \underline{H}$
4. The control ($\underline{AB}$ = 1) = trailing edge at $i = 2$ ($\underline{C}D$ = 1) causes transition $i = 2 \rightarrow 3$. (Note: To set the C-controlling flip-flop ON, trigger E, but not F. Do not reset the other flip-flop; i.e., do not trigger H.)	$\underline{A}\ \underline{B}\ \underline{C}\ D \rightarrow E\ \underline{F}\ \underline{H}$
5. The circuit remains steady at $i = 3$ (CD = 1) for idle control ($\underline{AB}$ = 1) and for SET ($A\underline{B}$ = 1). (No flip-flops to be reset.)	$\underline{A}\ C\ D \rightarrow \underline{F}\ \underline{H}$

Table 6.2 (Continued).

Formulation	Algebraic notation

6. The control $(A\underline{B} = 1)$ or $(AB = 1)$ at $i = 3$ $(CD = 1)$ causes the transition $i = 3 \rightarrow 1$.

    $A\ C\ D \rightarrow H\ \underline{G}\ \underline{F}$

7. The transition $i = 1 \rightarrow 0$ may proceed only when $(A\underline{B} = 1)$. For that reason, the circuit must remain steady at $i = 1$ $(C\underline{D} = 1)$ as long as $(A+B = 1)$.

    $(A + B)\ C\ \underline{D} \rightarrow \underline{F}\ \underline{G}$

8. The control $(A\underline{B} = 1)$ at $i = 1$ $(C\underline{D} = 1)$ causes the transition $i = 1 \rightarrow 0$.

    $\underline{A}\ \underline{B}\ C\ \underline{D} \rightarrow F\ \underline{E}\ \underline{G}$

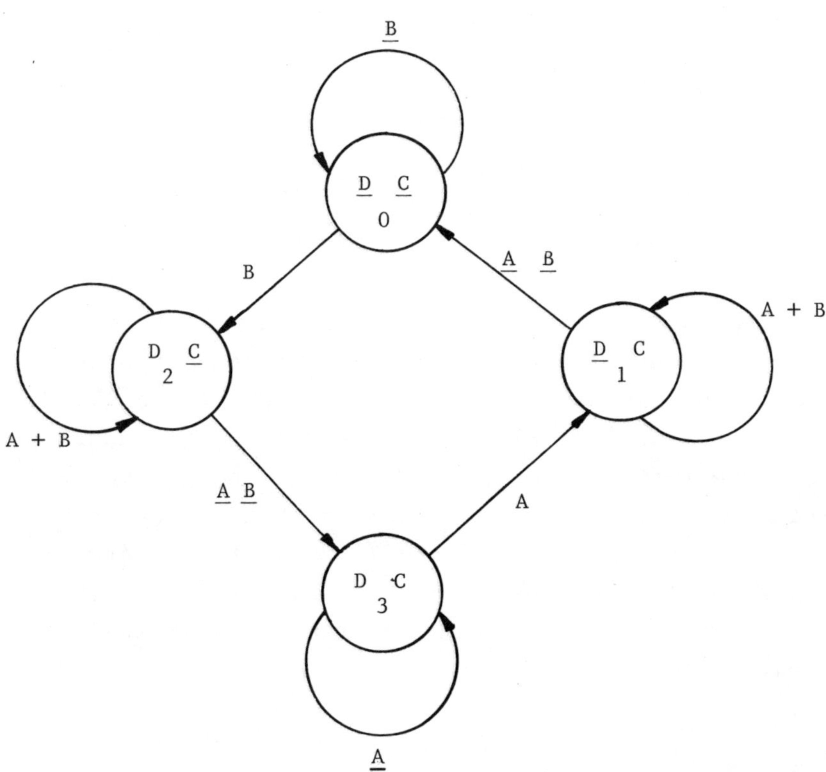

Fig. 6.11. State diagram
for constraints generation.

FORMULA
WRITE DOWN THE FORMULA:
C$\underline{A}$B$\underline{D}$→F$\underline{E}$G
MAY CALL EITHER (NEXT) FORMULA OR DISCRIMINANT OR EXECUTE.

The existence function of the system A, B, C, D, E, F, G, H is obtained by calling

DISCRIMINANT
THE DISCRIMINANT VALUE BEFORE CONSTRAINTS IS:
```
 1 1 0 0 0 1 1 1 0 1 1 1 1 0 1 0
 0 0 0 0 0 1 1 1 1 0 0 0 1 0 1 0
 1 1 0 0 1 0 0 0 0 1 1 1 0 0 0 0
 0 0 0 0 0 0 0 0 0 0 0 0 0 0 0 0
 0 0 1 1 0 0 0 0 0 1 1 1 1 0 1 0
 0 0 0 0 0 0 0 0 1 0 0 0 1 0 1 0
 0 0 1 1 0 0 0 0 0 1 1 1 0 0 0 0
 0 0 0 0 0 0 0 0 0 0 0 0 0 0 0 0
 1 1 0 0 0 1 1 1 0 0 0 0 0 1 0 1
 0 0 0 0 0 1 1 1 0 0 0 0 0 1 0 1
 1 1 0 0 1 0 0 0 0 0 0 0 0 0 0 0
 0 0 0 0 0 0 0 0 0 0 0 0 0 0 0 0
 0 0 0 0 0 0 0 0 0 0 0 0 0 0 0 0
 0 0 0 0 0 0 0 0 0 0 0 0 0 0 0 0
 0 0 0 0 0 0 0 0 0 0 0 0 0 0 0 0
 0 0 0 0 0 0 0 0 0 0 0 0 0 0 0 0
```
AFTER ::EXECUTE:: HAS BEEN CALLED, CALL MTY TO GET
THE CONSTRAINT RECTIFIED DISCRIMINANT.
CALL EXECUTE AFTER ALL CONDITIONS HAVE BEEN PUT IN.
IF NOT, CALL FORMULA AGAIN.

The procedure EXECUTE produces the outputs E, F, G, and H as functions of A, B, C, and D. The number of distinct solutions is extremely large (over 67 million). By typing 0 (zero) instead of a solution vector, the user can cause the terminal to print the solutions in such an order that the first solution contains the smallest number of minterm implicants and the last solution (here, solution number 67,108,864) contains that largest number of minterm implicants. This is important, as we shall demonstrate.

EXECUTE
NUMBER OF SOLUTIONS UNDER ALL CONSTRAINTS: SOL =   67108864
SOLUTION VECTOR OR TYPE 0 FOR CIRCUIT DESIGN.
□:
　　　0

SEQUENTIAL CIRCUIT AIDES

SOLUTION NUMBER: 1
E = [8] ∪ ()
F = [4] ∪ ()
G = [2   3] ∪ ()
H = [13   15] ∪ ()
SOLUTION NUMBER: 2
E = [8   15] ∪ ()
F = [4] ∪ ()
G = [2   3] ∪ ()
H = [13   15] ∪ ()
SOLUTION NUMBER: 3
E = [8   14] ∪ ()
F = [4] ∪ ()
G = [2   3] ∪ ()
H = [13   15] ∪ ()
SOLUTION NUMBER: 4
E = [8   14   15] ∪ ()
F = [4] ∪ ()
G = [2   3] ∪ ()
H = [13   15] ∪ ()
SOLUTION NUMBER: 5
E = [8] ∪ ()
F = [4] ∪ ()
G = [2   3   14] ∪ ()
H = [13   15] ∪ ()

    •

    •   (printout is interruptable)

    •

SOLUTION NUMBER: 67108864
E = [5   6   7   8   12   13   14   15] ∪ ()
F = [0   1   2   3   4   9   10   11] ∪ ()
G = [2   3   8   9   10   11   12   14] ∪ ()
H = [0   1   4   5   6   7   13   15] ∪ ()
SOLUTION NUMBER: 67108863
E = [5   6   7   8   12   13   14] ∪ ()
F = [0   1   2   3   4   9   10   11] ∪ ()
G = [2   3   8   9   10   11   12   14] ∪ ()
H = [0   1   4   5   6   7   13   15] ∪ ()
SOLUTION NUMBER: 67108862
E = [5   6   7   8   12   13   15] ∪ ()
F = [0   1   2   3   4   9   10   11] ∪ ()
G = [2   3   8   9   10   11   12   14] ∪ ()
H = [0   1   4   5   6   7   13   15] ∪ ()
SOLUTION NUMBER: 67108861
E = [5   6   7   8   12   13] ∪ ()
F = [0   1   2   3   4   9   10   11] ∪ ()
G = [2   3   8   9   10   11   12   14] ∪ ()
H = [0   1   4   5   6   7   13   15] ∪ ()
SOLUTION NUMBER: 67108860
E = [5   6   7   8   12   13   14   15] ∪ ()
F = [0   1   2   3   4   9   10   11] ∪ ()
G = [2   3   8   9   10   11   12] ∪ ()
H = [0   1   4   5   6   7   13   15] ∪ ()

IMPLEMENTATION OF THE RESULTS:

Version 1 -- without DON'T CAREs: The functions E, F,
G, and H are expressed algebraically. They are too simple
to require simplification by computer:

$$E = \underline{A}\ \underline{B}\ \underline{C}\ D; \quad F = \underline{A}\ \underline{B}\ C\ \underline{D}; \quad G = B\ \underline{C}\ \underline{D}; \quad H = A\ C\ D.$$

The dynamic schematic for this implementation is presented
in Fig. 6.12.

Version 2 -- with DON'T CAREs: By comparing solution
No. 1 with solution No. 67108864, we observe that each
minterm implicant of No. 1 is present in No. 67108864.
We conclude that minterms in No. 1 <u>must</u> be accepted as
minterm implicants; the additional minterms in No. 67108864
<u>may</u> by accepted (playing the role of DON'T CAREs).

To do the minimization by computer, call (in SYSTEM):

To get E: MINIMA 0 0 0 0 0 2 2 2 1 0 0 0 2 2 2 2
    (=> E = $\underline{A}\ \underline{B}$ D)

To get F: MINIMA 2 2 2 2 1 0 0 0 0 2 2 2 0 0 0 0
    (=> F = $\underline{A}\ \underline{B}\ \underline{D}$)

To get G: MINIMA 0 0 1 1 0 0 0 0 2 2 2 2 0 2 0
    (=> G = B $\underline{C}$)

To get H: MINIMA 2 2 0 0 2 2 2 2 0 0 0 0 0 1 0 1
    (=> H = A C)

The dynamic schematic implementing these equations is
shown in Fig. 6.13. Note that the use of DON'T CAREs
in the latter solution <u>simplifies</u> the design.

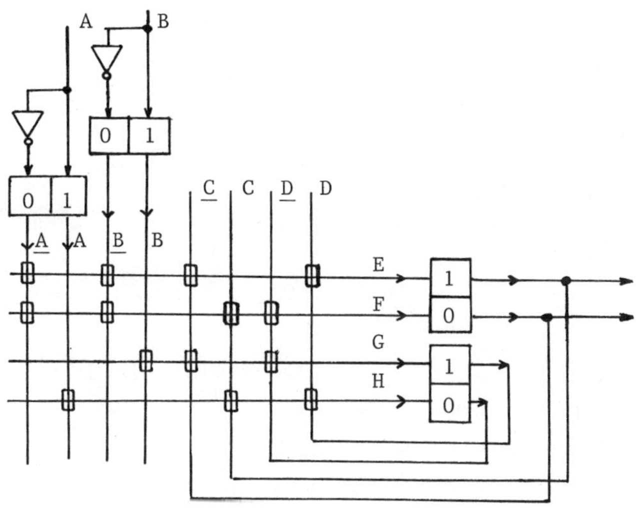

Fig. 6.12. J-K flip-flop with
trailing edge control.

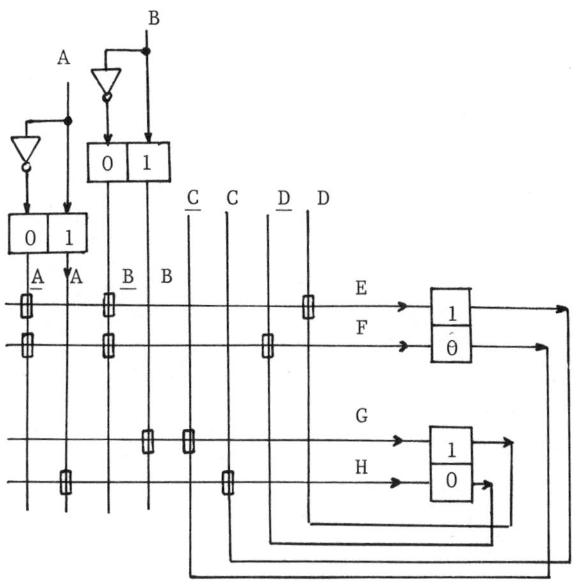

Fig. 6.13. Final design version of
J-K flip-flop.

**Chapter 7
The APL Program "BOOL"**

This chapter consists of the listings of the APL functions which comprise the program BOOL.

This concludes the APL program listings for the set of computer-assisted design programs developed by Dr. Svoboda.

THE APL PROGRAM "BOOL"

```
      ∇ BULL
[1]     'NUMBER OF CONSTANTS IS:'
[2]     NX←□
[3]     'SYMBOLS FOR CONSTANTS: ';ABCD[ιNX]
[4]     'NUMBER OF UNKNOWNS IS:'
[5]     NY←□
[6]     'SYMBOLS FOR UNKNOWNS: ';ABCD[NX+ιNY]
[7]     DUO←NX,NY
[8]     XXX←2*NX
[9]     YYY←2*NY
[10]    NN←2*N←NX+NY
[11]    TRFT←Nρ3
[12]    DSCR←(YYY,XXX)ρ1
[13]    'CALL FORMULA.'
      ∇


      ∇ COMB;BMB;P;U
[1]     BMB←AMB←2*(¯1+(ιN))
[2]     P←1
[3]   H1:CMB←ι0
[4]     U←P
[5]   H2:CMB←CMB,(((2*U)+(((2*U)>BMB)/BMB))
[6]     →(N>U←U+1)/H2
[7]     AMB←AMB,CMB
[8]     BMB←CMB
[9]     →(N>P←P+1)/H1
[10]    AMB←0,AMB
      ∇


      ∇ COMMENTS
[1]     'ENTER THE PROGRAM BY CALLING  BULL'
[2]     'NOT MORE THAN 12  VARIABLES'
[3]     'NEGATION BY UNDERLINING: A , B ,...'
[4]     '= STANDS FOR EQUALITY, → STANDS FOR IMPLICATION'
[5]     'VALUES:  0, 1  ARE PERMITTED ON THE RIGHT'
[6]     'SIDE OF THE FORMULA ONLY.'
[7]     'EXAMPLE OF AN EQUATION:  CDA+BA=BA+AC'
[8]     'THE SAME WITH SPACING: C D A + BA =...'
[9]     'EXAMPLE OF AN IMPLICATION: ABC+D→AD+CB'
[10]    'SOLUTION VECTOR = SET OF SOLUTION NUMBERS'
[11]    'ιSOL  WILL PRODUCE ALL SOLUTIONS'
[12]    'INTERPRETATION OF THE SOLUTION PRINTOUT:'
[13]    'EXAMPLE:  D = [1 4 5] ∪ (3 7)'
[14]    'MEANS: D  MUST BE TRUE FOR MINTERMS  1,4,5'
[15]    '        D  IS UNSPECIFIED FOR  3,7'
[16]    'CALL BULL' ≉CONST
      ∇
```

```
      ∇ CONST
[1]      ABCD←'ABCDEFGHJKLMABCDEFGHJKLM01→+ '
[2]      FABC←(3*(¯1+ι12)),(2×3*(¯1+ι12)),5ρ0
      ∇



      ∇ DISCRIMINANT
[1]      'THE DISCRIMINANT VALUE BEFORE CONSTRAINTS IS:   ';
         DSCR
[2]      'AFTER ::EXECUTE:: HAS BEEN CALLED, CALL MTY TO GET
         THE CONSTRAINT RECTIFIED DISCRIMINANT.'
[3]      'CALL EXECUTE AFTER ALL CONDITIONS HAVE BEEN PUT IN.'
[4]      'IF NOT, CALL FORMULA AGAIN.'
[5]      →0
      ∇



      ∇ EXECUTE;MTX;FIX;SET;COL;MMR;VEC;LIM;T;V;RES;MES;
         MAX;LIN;N;BMB;SEQ;SIG
[1]      MTX←⍉(XXX,YYY)ριYYY
[2]      SET←+/DSCR
[3]      DON←(YYY=SET)/¯1+ιXXX
[4]      NIC←(0=SET)/¯1+ιXXX
[5]      →(0=ρDON)/S5
[6]      'EACH SOLUTION HAS THE FOLLOWING DONT CARES:   ';
         DON
[7]   S5:→(0=ρNIC)/S6
[8]      'THE NUMBER OF SOLUTIONS IS ZERO, UNLESS
         THE FOLLOWING INPUT IDENTIFIERS ARE FORBIDDEN:   ';
         NIC
[9]   S6:MTY←(YYY,XXX)ρ((SET≠YYY)∧(SET≠0))
[10]     MTY←MTY×DSCR
[11]     MTY[1;]←(SET=0)∨(SET=YYY)∨MTY[1;]
[12]     MAX←⌈/SET←+/MTY
[13]     SOL←×/SET
[14]     MTX←MTX×MTY
[15]     'NUMBER OF SOLUTIONS UNDER ALL CONSTRAINTS:   SOL =   ';
         SOL
[16]     →(SOL=0)/0
[17]     'SOLUTION VECTOR OR TYPE  0   FOR CIRCUIT DESIGN.'
[18]     VEC←,⎕
[19]     →(0∈VEC)/L1
[20]  L3:LIM←ρVEC
[21]     →L9
[22]  L1:→(SOL≥LIM←2×1+NY)/L2
[23]     VEC←ιSOL
[24]     →L3
[25]  L2:VEC←(ι(1+NY)),1+SOL-ι(1+NY)
[26]  L9:MES←(MAX,XXX)ρ0
[27]     V←1
```

```
[28]    N←NY
[29]    COMB
[30] L4:COL←(MTX[;V]>0)/MTX[;V]
[31]    BMB←1+AMB
[32]    SIG←ρSEQ←BMBιCOL
[33]    SEQ←BMB[SEQ[⍋SEQ]]
[34]    LIN←MAXρ0
[35]    LIN[ιSIG]←SEQ
[36]    MES[;V]←LIN
[37]    →(XXX≥V←V+1)/L4
[38]    T←1
[39] L5:FIX←1+SETτ(VEC[T]-1)
[40]    RES←ι0
[41]    V←1
[42] L6:COL←MES[;V]
[43]    RES←RES,COL[FIX[V]]
[44]    →(XXX≥V←V+1)/L6
[45]    MMR←(NY,XXX)ρ0
[46]    RES←RES-1
[47]    V←1
[48] L7:MMR[;V]←V×(NYρ2)τRES[V]
[49]    →(XXX≥V←V+1)/L7
[50]    MMR←⊖MMR
[51]    'SOLUTION NUMBER: ';VEC[T]
[52]    V←1
[53] L8:ABCD[NX+V];' = [';¯1+((0<MMR[V;])/MMR[V;]);
        '] ∪ (';DON,NIC;')'
[54]    →(NY≥V←V+1)/L8
[55]    →(LIM≥T←T+1)/L5
     ∇
```

```
     ∇ FORMULA;FCT;RES;SETL;SETR;LEF;RIG;PUL;FORM;ALL
[1]    'WRITE DOWN THE FORMULA:'
[2]    ALL←ρFORM←⎕
[3]    PUL←⌊/(FORMι'='),(FORMι'→')
[4]    →(0=ρLEF←FORM[ι(PUL-1)])/0
[5]    →(0=ρRIG←FORM[PUL+ι(ALL-PUL)])/0
[6]    SETL←REFORM LEF
[7]    SETR←REFORM RIG
[8]    →('→'∊FORM)/F1
[9]    RES←((¯1+ιNN)∊SETL)=((¯1+ιNN)∊SETR)
[10]   →F2
[11] F1:RES←((¯1+ιNN)∊SETL)≤((¯1+ιNN)∊SETR)
[12] F2:FCT←(YYY,XXX)ρRES
[13]   DSCR←DSCR×FCT
[14]   'MAY CALL EITHER (NEXT) FORMULA OR DISCRIMINANT
        OR EXECUTE.'
[15]   →0
     ∇
```

```
      ∇ REF←REFORM FRM;PUT;ORG;U;TRI;TRFD;TRM;MEZ;RES;S
[1]     REF←ι0
[2]     →('0'∈FRM)/0
[3]     →('1'∈FRM)/R2
[4]     PUT←(FRM='+')/ιρFRM
[5]     ORG←0,PUT,ρFRM
[6]     U← 1 2
[7]   R1:TRI←+/FABC[ABCDι(FRM[ORG[U[1]]]+ι(ORG[U[2]]-
        ORG[U[1]])])]
[8]     TRM←TRFTτTRI
[9]     TRFD←1+(TRM=0)
[10]    MEZ←2*+/TRM=0
[11]    S←0
[12]  R4:RES←2|TRM+TRFDτS
[13]    REF←REF,2⊥RES
[14]    →(MEZ>S←S+1)/R4
[15]    →((ρORG)≥⌈/U←U+1)/R1
[16]    →0
[17]  R2:REF←¯1+ιNN
      ∇
```

8.1   INTRODUCTION

The emphasis on clean, irredundant, minimal designs
has been dramatically affected by the evolution of LSI
technology.   There are instances where a minimal implementation costs more in board space or design time without providing testing, cost, or execution speed advantages.   An understanding of minimization techniques is
still a requirement for the LSI designer, however, as
much of the logical thought discipline is necessary for
current micro system designs.

For the chip designer, circuit cost can be shown
to be linearly related to chip area, which in turn is a
function (all other variables remaining fixed) of the
number of gates and the number of connections.   The
emphasis of minimization would be gate count reduction
for TTL and ECL (bipolar LSI), since gates occupy a
larger chip area than connections; for MOS LSI, the
emphasis of the minimization would be connection length
or connection count (the connection length is related to
the number of gates and to the number of connections).

Although highly simplified, expressing circuit cost
as a function of the number of gates and the number of
connections gives a reasonable approximation.

For the circuit board designer, cost can be shown
to be related to the number of packages and the number
of connections.   For high-speed designs, the emphasis of

the minimization would be connection length, which is a function of the number of gates and the number of connections. For densely packed boards, the emphasis of the minimization would be package count. It should also be noted that, for a fixed board size, as the cost of the IC packages drops, the cost of the connections becomes dominant and the emphasis of the minimization shifts accordingly.

For testing purposes, redundant circuits are not as testable as their irredundant equivalents, since redundancy masks faults (see Chapter 12).

From the preceding, it is evident that judicious minimization remains an important part of any cost-effective design and implementation.

This and the following chapters will present some of the new techniques in minimization such as Svoboda's weight algorithm and the fundamental product concept, and some of the logical instrument teaching aides that have been used to demonstrate the theorems behind these techniques.

8.2 SINGLE-OUTPUT MINIMIZATION

For simplification, the minimization techniques will be presented using a single-output combinational circuit as shown in Figure 8.1. (A brief look at multiple-output minimization is included in Section 8.6.)

8.2.1 Design Constraints

The selection of a minimization technique is a function of the overall objectives. Any design must be implemented under certain constraints similar to those called out in Table 8.1. A design is also broken up into functional modules prior to implementation.

The need to partition a module further into units that are of a reasonable size for human comprehension and/ or for the purpose of honoring the limits of design support systems will restrict the actual "cleanness" of the final design. By breaking a circuit up into modules, some opportunities for minimization will become forbidden. The further partitioning into submodules will also affect the minimization. The tradeoff is the design time. Each additional variable increases the problem complexity by a factor of two in the binary space. Therefore, it is more efficient if, after the initial module design is completed, it is reviewed for further reductions.

Sub-module size is a function of the techniques available for the design. If an APL program such as the package in the first seven chapters of this book is used, the restriction is thirteen to fifteen variables (limited by the workspace size). The parallel Boolean processor, if built as discussed in a later chapter, could handle up to 22-variable modules.

The design constraints such as device type, board space, etc., will determine the desired result of the minimization to be used. For example, NOR-NOR gate staging implements the minimal $\Pi\Sigma$-form, while NAND-NAND gate staging implements the minimal $\Sigma\Pi$-form. If multiplexors may be used, Marquand mapping and column-only minimization should be performed.

For any of these approaches, the underlying minimization techniques are the same.

8.2.2. Minimization by Inspection

There are many instances where the algebraic expression or the logical map (Venn, Veitch, Karnaugh, Marquand, Triadic) is simple enough for obvious reductions to be made without any particular technique being formally applied. This is the case for simple expressions of six or fewer variables such as

$$Y = A + AC + A\underline{B}$$

which may be immediately rewritten as

COST	Affects everything
FAN-IN, FAN-OUT	Function of logic family to be used
LOGIC TYPE	Any constraint on NAND, NOR, etc., availability
TIMING CONSIDERATIONS	High-speed logic requirement
BOARD SIZE	Available "real estate" (restricts package/gate count)
RELIABILITY REQUIREMENTS	Constrains redundancy allowed
TIME	How much effort can be expended
SUPPORT	Computer assist; manual reductions
POWER REQUIREMENTS	Further constraints on implementation
NUMBER OF VARIABLES	Pinout limitations (primary variables; internal or secondary variables are a function of the number of connections)
PARTITIONING	Breaking a module up into manageable pieces
MODULE DEFINITION	Module boundaries should be maintained
CLASSIFICATION OF DESIGN	Spacecraft (most minimal, most reliable); military (rugged); commercial

Table 8.1. Design Constraints.

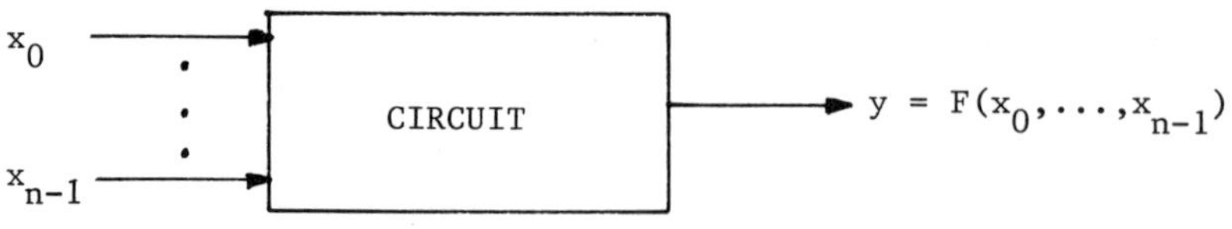

Figure 8.1. Combinational, single-output circuit.

$$Y = A$$

by recognizing that

$$A (1 + C) = A (1) = A.$$

An equivalent example for a map would be the appearance of an obvious structure as shown in Figure 8.2 for Karnuagh and Marquand maps of

$$y = \overline{x}_3\overline{x}_1 + \overline{x}_3x_2 + x_2x_1.$$

8.2.3 Minimization by Mapping and Observation

For a small number of variables, a function to be minimized may be mapped by expanding the expression into a sum of products form and marking a "one" on the map at all points corresponding to a minterm of the function. Points which are logical distance one apart are connected. (On a Karnaugh 4-variable map, these would be adjacent points.) The resulting structures represent reduced terms. The terms of the largest structures form the prime implicants of the function.

This is a casual approach and as the number of variables and/or the number of minterms of the function increases, the reliability of this method decreases.

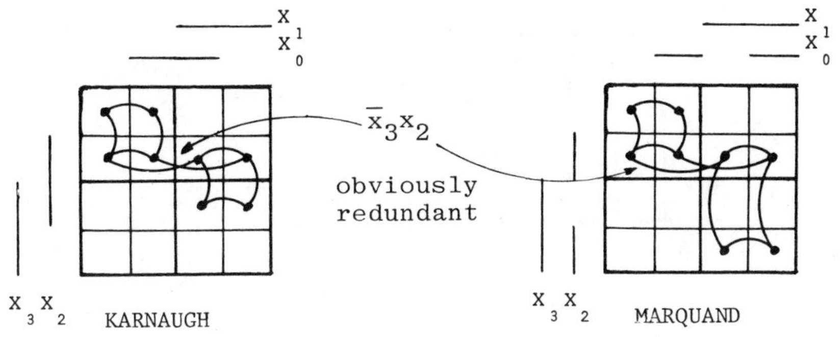

Figure 8.2. Reducing a simple function.

8.2.4 Minimization by Algebraic Manipulation

Algebraic manipulation of an expression for a function into a reduced minimal form is tedious and prone to human error. The same disadvantages may be cited for tabular reduction techniques.

Figure 8.3 demonstrates an algebraic reduction and presents a Marquand map of the sample function.

$$Y = BDE + \bar{B}\bar{C}D + CDE + \bar{A}\bar{B}CE + \bar{A}\bar{B}C + \bar{B}\bar{C}\bar{D}\bar{E}$$

$$= BDE + \bar{B}\bar{C}D + CDE + \bar{A}\bar{B}(CE + C) + \bar{B}\bar{C}\bar{D}\bar{E}$$

$$= BDE + \bar{B}\bar{C}D + CDE + \bar{A}\bar{B}C + \bar{B}\bar{C}\bar{D}\bar{E}$$

$$= BCDE + B\bar{C}DE + \bar{B}CDE + BCDE + \bar{B}\bar{C}DE + \bar{B}\bar{C}D\bar{E} + \bar{A}\bar{B}C + \bar{B}\bar{C}\bar{D}\bar{E}$$

$$= BCDE + B\bar{C}DE + \bar{B}CDE + BCDE + \bar{B}\bar{C}DE + \bar{B}C\bar{E} + \bar{A}\bar{B}C$$

$$= BDE(C + \bar{C}) + \bar{B}DE(C + \bar{C}) + \bar{B}C\bar{E} + \bar{A}\bar{B}C$$

$$= DE(B + \bar{B}) + \bar{B}C\bar{E} + \bar{A}\bar{B}C$$

$$= DE + \bar{B}\bar{C}\bar{E} + \bar{A}\bar{B}C$$

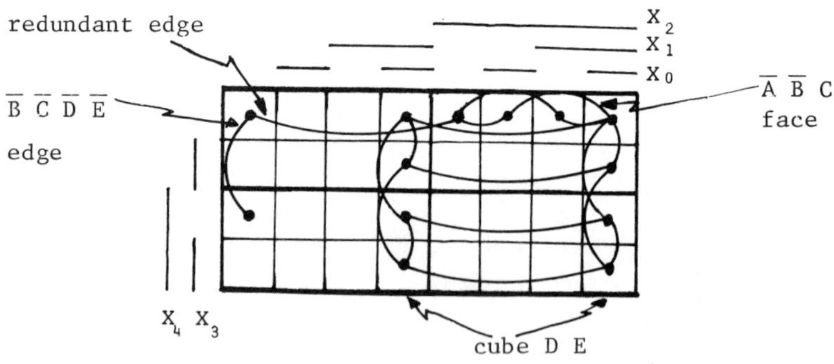

Figure 8.3. Algebraic and map minimization
of a five-variable function.

8.3. Svoboda's Weight Algorithm

A convenient algorithm for manual or programmed minimization of a function is the weight algorithm developed by Svoboda. It is readily applied to manual solutions of up to eight variables depending upon the complexity of the function. To perform the weight algorithm manually, proceed as follows:

1. Map all terms where the function Y is true on a Marquand map as "1" points. (A Karnaugh map may be used, but it is inconvenient.) Include all "don't care" terms as "#" points. Unmarked points are those for which the function Y = 0.

2. Connect all pairs of points of logical distance one where p_i = 1 or p_i = # where p_i is the label of the point i. All such connections are referred to as "edges".

3. Using a second map (for clarity), fill in the squares corresponding to the minterms of the function Y (all p_i = 1) with the number of edges connected to that minterm. Include in the count edges between minterms where both p_i = 1 and between minterms where one p_i = #.

4. Scanning the points sequentially from the origin (p_0), find that minterm with the lowest edge count or "weight". (The first search should be for points with weight w = 0.) This is a CRITICAL POINT of the function Y.

   In his paper "Ordering of Implicants", Svoboda discusses the natural phenomena that, when coverage is made from the origin forward in sequence, the probability of the minimal function expression being obtained is increased. The ordering concept is the basis of the weight algorithm, fundamental product coverage, and the multiple-output minimization.

5. Select the term representing the largest structure (edge, face, cube) which covers that minterm. This is a PRIME IMPLICANT of the function.

6. Record the prime implicant and mark all minterms of the prime implicant as "covered" by labelling the points as "don't cares".

7. Continue scanning from the last critical point to find the next uncovered minterm with the lowest weight. If there are none with the last-selected weight value, increment the weight value by one and return to the origin to begin scanning again.

8. Repeat steps 5 through 7 until all minterms of the function are covered. The selected minterms form the critical set of minterms of the function.

The problem with the algorithm is that where a choice of structures exists, the algorithm fails. In some cases a solution may be obtained by choosing the structure which covers the most 1's or "uncovered" minterms. Where there are equal choices, there is more than one solution and the algorithm fails to provide a choice of one best solution. In many cases, there is little practical advantage in pursuing more than one of the minimal solutions.

The algorithm is presented here without proof. Theorems for this and other procedures are presented in Chapter 9.

Figure 8.4 presents a step-by-step solution of a five-variable function using Marquand maps. Figure 8.5 presents an interesting six-variable, incompletely specified function. The latter example appears throughout the text.

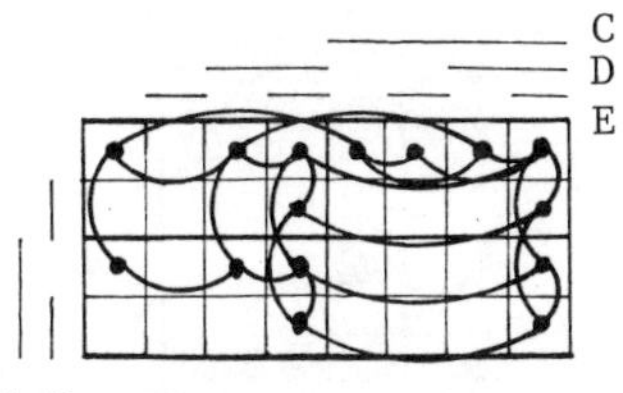

a. Edges of the function

$$y = BDE + \overline{B}\overline{C}D + CDE$$
$$+ \overline{A}BCE + \overline{A}\overline{B}C + \overline{B}\overline{C}\overline{D}E$$

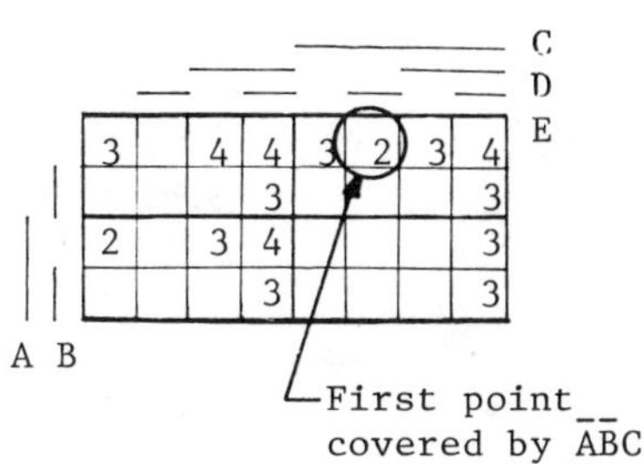

b. Weights of the function, showing the first critical point with W = 2.

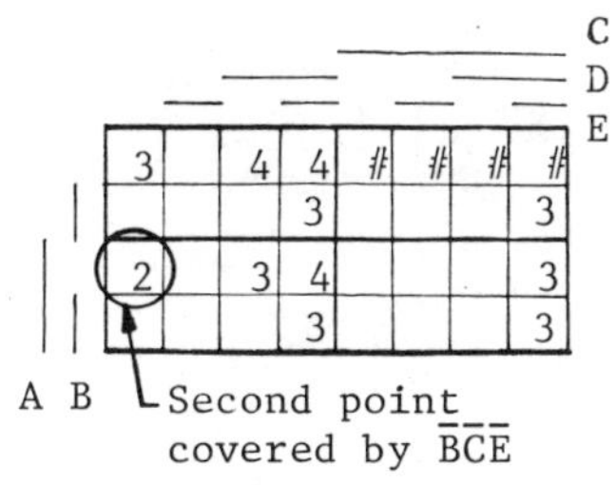

c. The second critical point with W = 2.

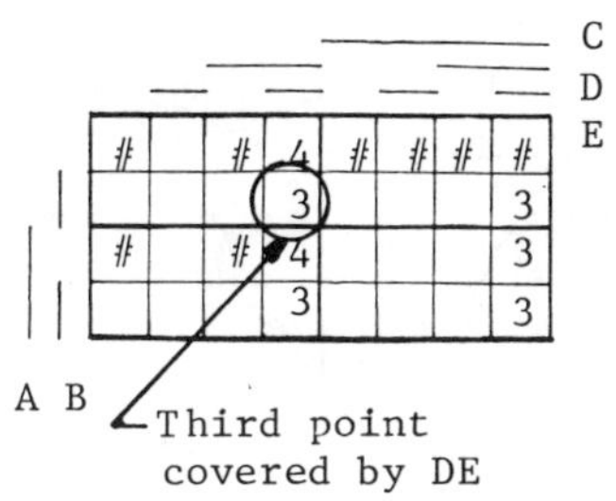

d. The third critical point with W = 3

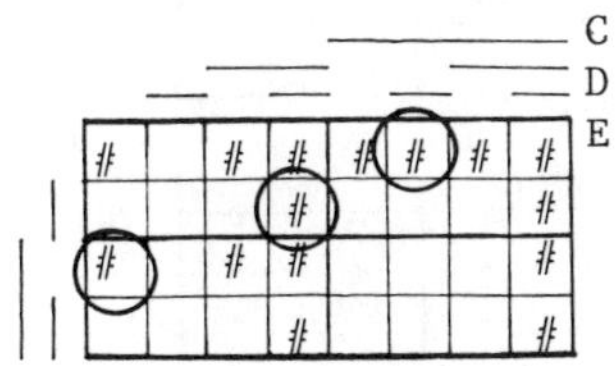

Coverage is complete

e. Final function is

$$y = \overline{A}\overline{B}C + \overline{B}C\overline{E} + DE$$

◯ indicates critical point

Figure 8.4. The weight algorithm for a five-variable function.

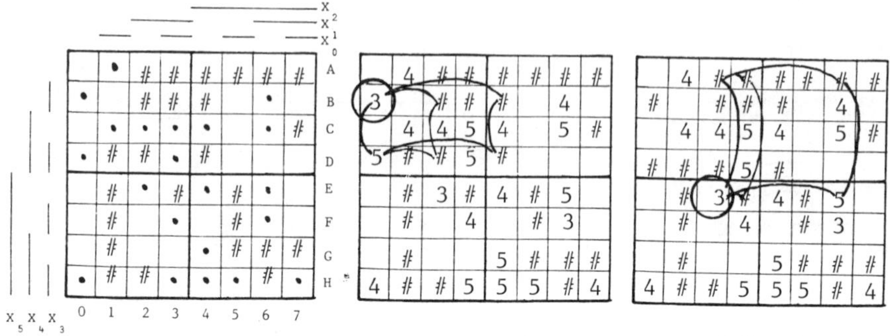

a. Initial map of the   b. Choose: $\bar{x}_5 x_3 \bar{x}_1 \bar{x}_0$.   c. $\bar{x}_4 \bar{x}_3 x_1 \bar{x}_0$.
    function

Maps b - i show the weights of the uncovered
minterms. For each term, choose the one which
covers the most uncovered minterms.

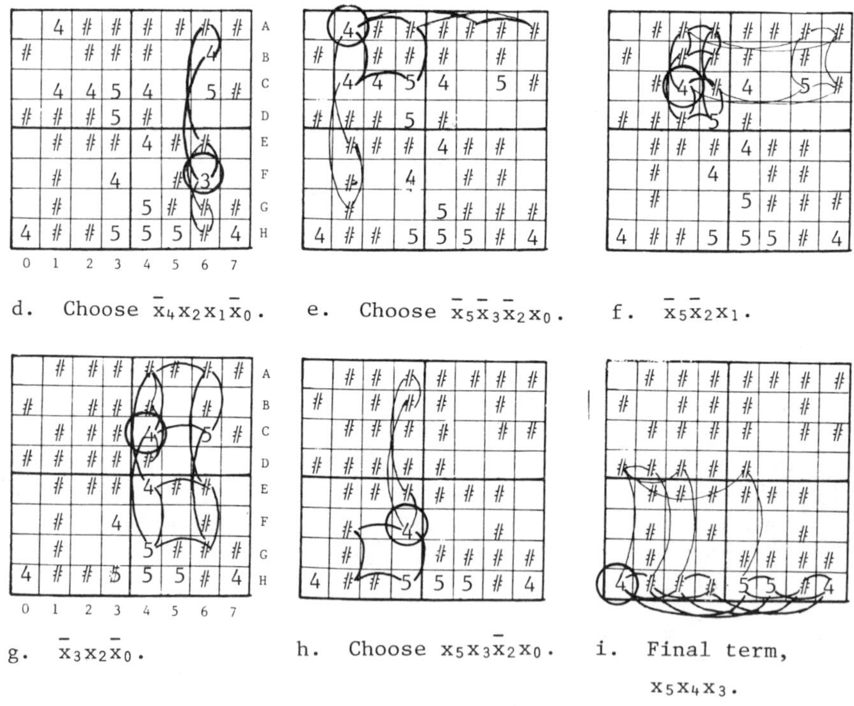

d. Choose $\bar{x}_4 x_2 x_1 \bar{x}_0$.   e. Choose $\bar{x}_5 \bar{x}_3 \bar{x}_2 x_0$.   f. $\bar{x}_5 \bar{x}_2 x_1$.

g. $\bar{x}_3 x_2 \bar{x}_0$.   h. Choose $x_5 x_3 \bar{x}_2 x_0$.   i. Final term,
                                                            $x_5 x_4 x_3$.

Figure 8.5. The weight algorithm
for a six-variable function.

8.4 LOGICAL INSTRUMENTS: THE WEIGHT DECK

8.4.1 Description of the Cards

As a teaching aid for the weight algorithm, Svoboda developed a deck referred to as the "weight deck", a set of 80-column punched cards where each card represents a point in the six-variable binary space.

Each card is indexed in the upper left-hand corner by the row-column short-hand index in the form "xn" (letter-digit). Each card is also indexed in the upper right corner by the decimal point identifier.

The card is punched with a group of six vertically-grouped punches for each point in the binary space. Holding the card so that the eighty columns run vertically from top to bottom and so that the punch rows zero through seven run from left to right, the groups are seen to be arranged in rows and columns corresponding to the Marquand map (see Figure 8.6).

A "sextet" is nonpunched at the point corresponding to the point the card represents. One punch in a sextet is nonpunched if the point that sextet represents is at logical distance one from the point represented by the card. The particular punch position corresponds to the variable which changes (the variable missing from the term representing the edge formed by the two points).

8.4.2. Finding the Weights for the Six-Variable Example

Referring to Figure 8.5a, to obtain the weights or edge counts for the minterms of the function, from the set of 64 cards representing the binary space, remove those cards which represent the zeros of the function.

The set of cards representing the zeros, when aligned and held to a light for observation, will show one punched position in each group of punch positions for each edge which may be formed between "1" and "don't care" points of the function.

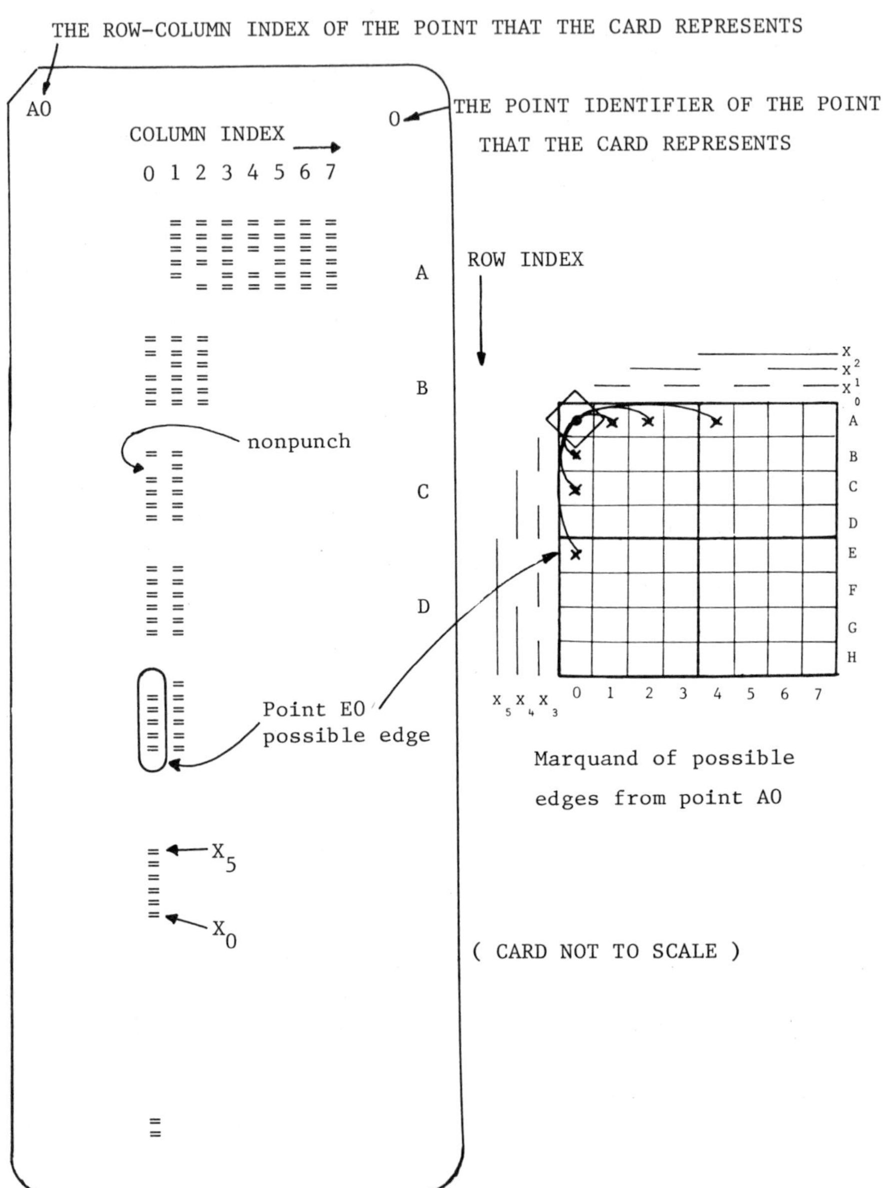

Fig. 8.6. Weight deck card A0.

By reading a count of the punch positions punched in each sextet, the weight of the point that sextet represents is obtained.

In addition, the punch position contains the information describing which edge of the six possible for any point exists. The lowest punch position of the sextet corresponds to x_0 and the highest to x_5.

For our example, use the cards representing the points: A0, B1, B5, B7, C0, C5, D5, D6, D7, E0, E7, F0, F2, F4, F7, G0, and G2. These cards will produce the weights for the "1" and for the "don't care" points. (The maps shown earlier did not record the weights for the don't cares -- see Figure 8.5b.)

8.5   SVOBODA'S FUNDAMENTAL PRODUCT PROCEDURE

8.5.1   Introduction

For more difficult problems, the structures or terms which cover a given minterm may not be "visible" or more than one structure of seemingly identical properties may cover the minterm. In these cases, the choice of the proper term for the best coverage becomes non-trivial. The fundamental product and the theorems of mutual term exclusivity were developed for these situations.

The fundamental product of a minterm is defined as the product of those literals which occur in every edge which can be connected to the minterm or point under examination. In other words, it is the product of those literals which are present in every term which covers the minterm.

An EFFECTIVE LITERAL is a literal that, when added to the literals of a fundamental product, forms a product that causes ones to be included in the term and no zeros to be included.

Any minterm which is also covered by <u>all</u> of the prime implicants which are found for the minterm under examination is called MUTUALLY TERM INCLUSIVE.

8.5.2. The Procedure for Finding the Fundamental Product and the Effective Mask

To manually perform minimization via the fundamental product modification to the weight algorithm, proceed as follows (note that this is a programmable procedure):

1. Perform the first steps of the weight algorithm to find the weights of all the minterms of the function to be reduced.

2. Begin at the origin and scan in decimal index order to find the first critical point (as per the weight algorithm).

3. Find the fundamental product for that minterm by listing <u>all</u> of the edges that can be formed from that minterm to points which are minterms or points which are don't cares. The fundamental product will be those literals which appear in each and every edge.

4. Examine (using a scratch map) only those points represented by the fundamental product. Are there any literals which may be added to the fundamental product to form an effective mask such that all ones visible on the scratch map are covered, but no zeros are covered? These literals form the effective mask.

    If there is a failure to find an effective mask, mark this point as "failed" and proceed to the next critical point and repeat the process.

    If there is success, the term so formed belongs to the minimal form. Mark all minterms which have been covered by this term as "don't cares" and proceed to the next critical point.

5. Periodically retry "failed" points.

6. Repeat steps 3 through 5 until all minterm points are covered.

Part of the solution of the six-variable problem using the fundamental product approach is shown in Figure 8.7. The partial solution should be compared to that of Figure 8.5. Note that it is still possible to have multiple solutions.

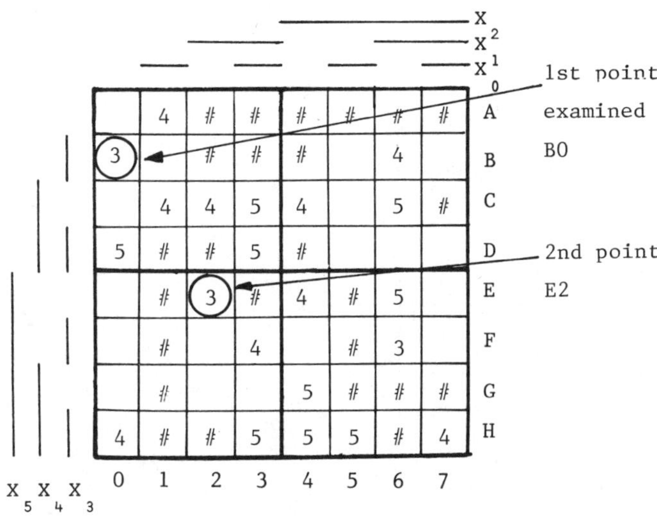

Figure 8.7. Fundamental product solution.

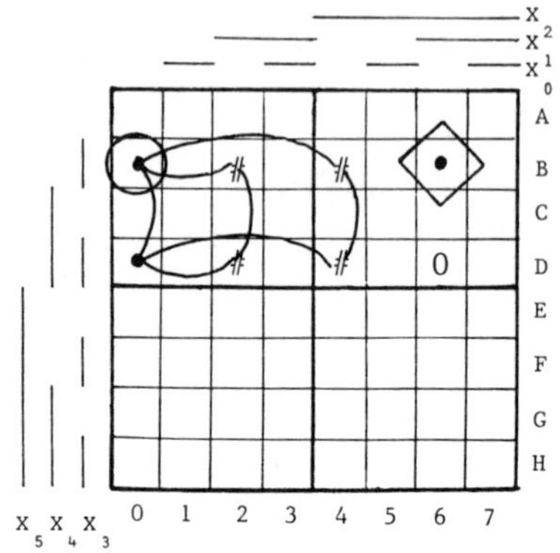

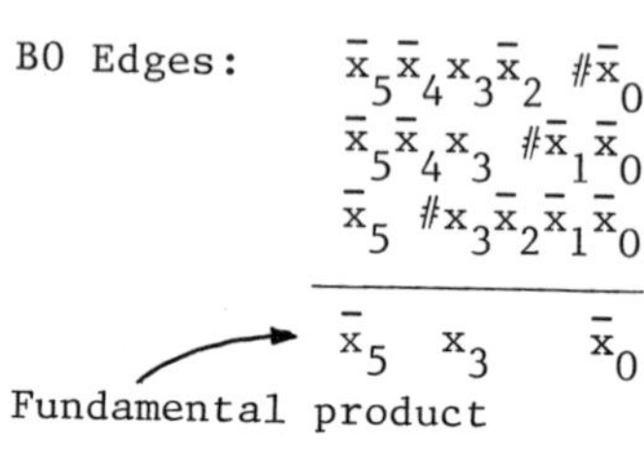

B0 Edges:

$$\bar{x}_5\bar{x}_4 x_3 \bar{x}_2 \not\!\bar{x}_0$$
$$\bar{x}_5\bar{x}_4 x_3 \not\!\bar{x}_1 \bar{x}_0$$
$$\bar{x}_5 \not\! x_3 \bar{x}_2 \bar{x}_1 \bar{x}_0$$

$$\bar{x}_5 \quad x_3 \qquad \bar{x}_0$$

Fundamental product

No effective mask   <u>FAILURE</u>

$\bar{x}_5 x_3 \bar{x}_2 \bar{x}_0$ leaves B6 uncovered
$\bar{x}_5 x_3 \bar{x}_1 \bar{x}_0$ leaves B6 uncovered

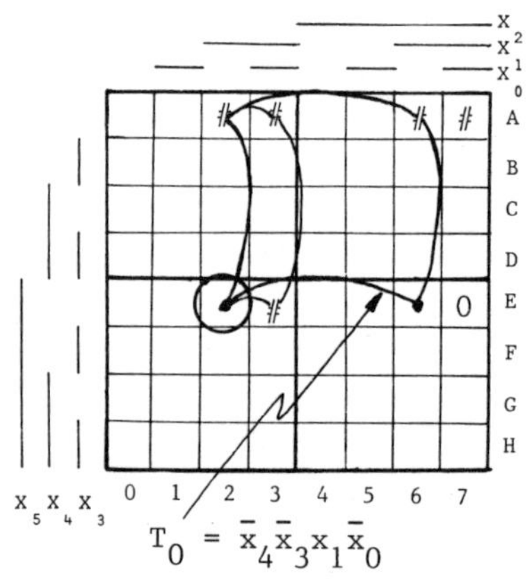

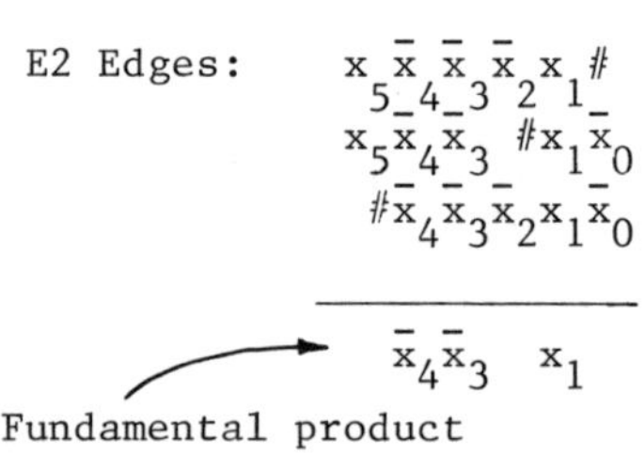

E2 Edges:

$$x_5 \bar{x}_4 \bar{x}_3 \bar{x}_2 x_1 \not\!$$
$$x_5 x_4 x_3 \not\! x_1 \bar{x}_0$$
$$\not\! \bar{x}_4 x_3 \bar{x}_2 x_1 \bar{x}_0$$

$$\bar{x}_4 \bar{x}_3 \quad x_1$$

Fundamental product

Effective mask: $\bar{x}_4\bar{x}_3 x_1 \bar{x}_0$
covers all ones, no zeros
($\bar{x}_4\bar{x}_3\bar{x}_2 x_1$ leaves E6 uncovered)

<u>SUCCESS</u>

Figure 8.7 (cont.).

4th point, A1

A1: Fundamental product: $\bar{x}_3 x_0$
Effective mask: $\bar{x}_5 \bar{x}_3 \bar{x}_2 x_0$
SUCCESS

3rd point,
FAILURE

$x_5 x_4 x_3$ 0 1 2 3 4 5 6 7

6th point, FAILURE
5th point, FAILURE
7th point, FAILURE
8th point, SUCCESS
E4: Effective mask: $\bar{x}_3 x_2 \bar{x}_0$

NOTE

$x_5 x_4 x_3$ 0 1 2 3 4 5 6 7

Retry F6 (3rd point)
Fundamental product
$x_2 x_1 \bar{x}_0$
Effective mask
$\bar{x}_4 x_2 x_1 \bar{x}_0$
SUCCESS

$T_3 = \bar{x}_4 x_2 x_1 \bar{x}_0$

$x_5 x_4 x_3$ 0 1 2 3 4 5 6 7

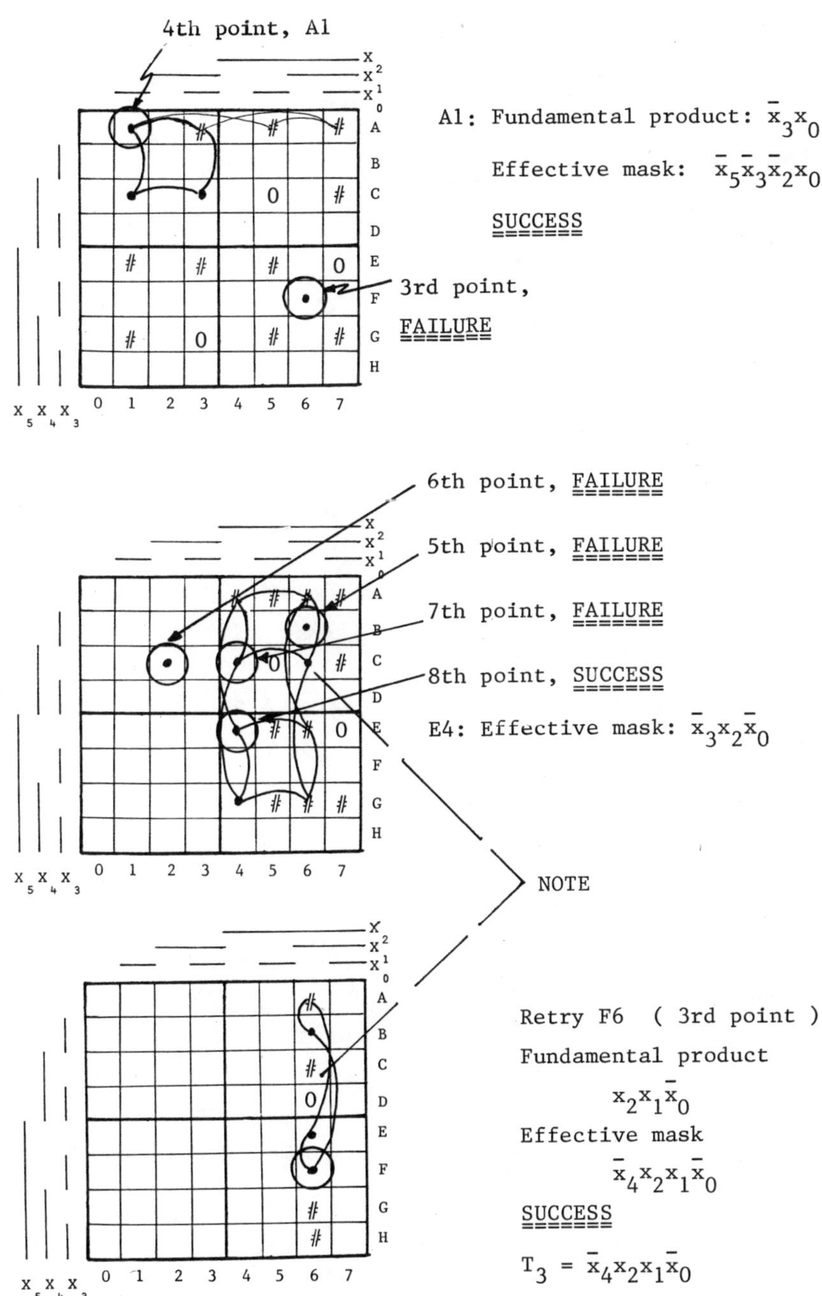

Fig. 8.7. (cont.).

Chapter 9
Introduction to Triadic Graphical Calculus

9.1   INTRODUCTION

This chapter introduces the binary and triadic notations which appear in the referenced papers on the parallel Boolean processor and mosaic functions.

Included is the triadic map algorithm for finding
the complete sum of the function $y = F(x)$, which is part
of the set of applications for the parallel Boolean processor. This algorithm is programmable and is based upon the
theorems of Chapter 11. The chapter concludes with the
triadic and binary logical instrument card decks which
have been successfully used to demonstrate the algorithm.

9.2   BINARY SPACE NOTATION

A system X of n Boolean variables where

$$x_i \in \{0,1\} \text{ for } 0 \leq i \leq n-1$$

may be represented in terms of the ALGEBRA OF SETS by a
LOGICAL SPACE which is the universe of the system.

The logical space consists of all points x where
each point represents a particular validity configuration
of the variables of the system. The total number of
points in the logical space is 2^n. The Marquand map will
be used to represent the logical space.

The point x in the logical space is identified by:

$$x = (x_{n-1} \, x_{n-2} \, \cdots \, x_1 \, x_0)_2$$
$$= x_{n-1} \cdot 2^{n-1} + x_{n-2} \cdot 2^{n-2} + \ldots + x_1 \cdot 2^1 + x_0 \cdot 2^0$$

INTRODUCTION TO TRIADIC GRAPHICAL CALCULUS

The integer x is termed a POINT IDENTIFIER.

A MINTERM in Boolean algebra is a function defined by:

$$m_a = \dot{x}_{n-1} \; \dot{x}_{n-2} \; \cdots \; \dot{x}_1 \; \dot{x}_0$$

where

$$\dot{x}_i \in \{x_i, \bar{x}_i\} \quad \text{for } 0 \le i \le n-1$$

The function m_a is equal to 0 everywhere except for a single configuration of values for the variables. When this function is charted in the logical space, the space is filled with zeros everywhere except at a single point where it is 1.

It has been an accepted policy to give the index of m_a the value of the point identifier for the point where $m_a = 1$. Under that convention:

$$m_a = 0 \text{ for } x \ne a$$

$$m_a = 1 \text{ for } x = a$$

The index a is termed the MINTERM IDENTIFIER and has the form of a binary number

$$a = a_{n-1} \cdot 2^{n-1} + a_{n-2} \cdot 2^{n-2} + \ldots + a_1 \cdot 2^1 + a_0 \cdot 2^0$$

where

$$a_j \in \{0,1\}$$

The following is the correspondence between a and the minterm identified:

$\dot{x}_j$	x_j	$\bar{x}_j$
a_j	1	0

Or:

$$(\dot{x}_j = x_j) \iff (a_j = 1)$$

$$(\dot{x}_j = \bar{x}_j) \iff (a_j = 0)$$

Figure 9.1 demonstrates the Marquand map for $m_a = m_{10}$.

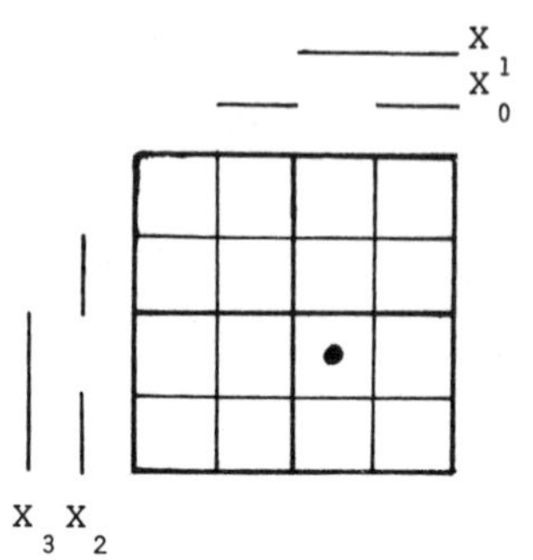

Figure 9.1.   Marquand map for the
function $m_a = x_3\overline{x_2}x_1\overline{x_0}$ ($m_a = m_{10}$).

9.3   TRIADIC SPACE NOTATION

A TERM is a Boolean function defined by:

$$t_h = t_h(x) = \ddot{x}_{n-1}\ \ddot{x}_{n-2}\ \cdots\ \ddot{x}_1\ \ddot{x}_0$$

where

$$\ddot{x}_i \in \{1,\ x_i,\ \overline{x_i}\}$$

For a logical system X with n Boolean variables, Svoboda created a TRIADIC SPACE of 3^n points in order to represent all possible terms belonging to the logical system.   A triadic map is used to represent the triadic space.

The TERM IDENTIFIER h is defined as a triadic number:

$$h = h_{n-1}\cdot 3^{n-1} + h_{n-2}\cdot 3^{n-2} + \ldots + h_1\cdot 3^1 + h_0\cdot 3^0$$

where

$$h_i \in \{0,1,2\}$$

The following is the correspondence between h_j and the term identified:

$\ddot{x}_j$	1	x_j	$\overline{x_j}$
h_j	0	1	2

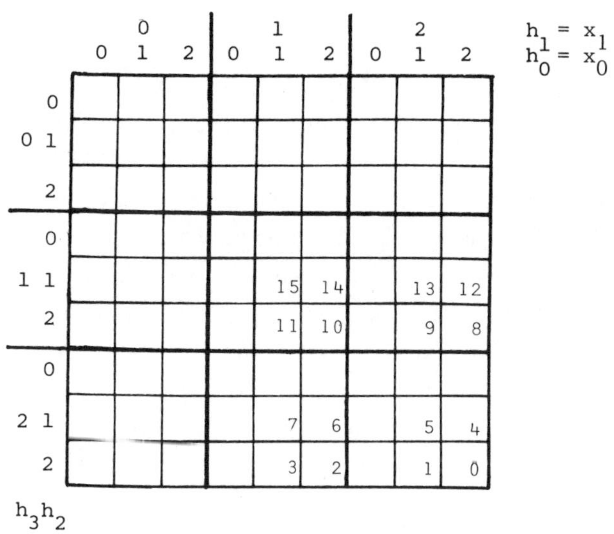

Figure 9.2. Triadic map with minterm space
labeled with point identifiers.

Or:

$$(\ddot{x}_j = 1) \iff (h_j = 0)$$

$$(\ddot{x}_j = x_j) \iff (h_j = 1)$$

$$(\ddot{x}_j = \overline{x}_j) \iff (h_j = 2)$$

The triadic map was devloped by Svoboda as a graphical medium for the representation of Boolean forms. A
triadic map for four-variable space is shown in Figure 9.2.
Note that the space of minterms (or minterm cube) is a
subspace of the triadic space. The minterm cube on the
triadic map includes only those terms t_h which have only
nonzero digits as the coordinates of their identifiers.

Or,

$$\text{Space of minterms} = \{t_h \mid h_j \in \{1,2\}\}$$

9.4 CORRELATIONS BETWEEN BINARY AND TRIADIC SPACE

The Karnaugh map, Marquand map, and triadic map
for a four-variable system are shown in Figure 9.3.
Each is labeled with point identifiers. Figure 9.3c
demonstrates the relationship between triadic terms
53 and 80 and minterms 8 and 0 respectively.

Figure 9.4 demonstrates the relationships between
the terms of the triadic space and the configurations
of the binary space for four variables using the Marquand
map. Note that the larger structures have the lower-
numbered point identifier. This is an important feature
of the triadic map and relates to the "ordering of implicants" referred to previously.

9.5 TRIADS

A TRIAD is a set of three points:

$$\{t_h \mid h \in \{a,b,c\}\}$$

from the triadic space whose identifier coordinates h_i
agree for all variables except one, x_j. For the variable
x_j, the coordinates of the points of a triad are bound
through the relationships:

h	a	b	c
h_j	0	1	2

The terms of the points of any triad are logically related
through:

$$t_c = \bar{x}_j t_a$$
$$t_b = x_j t_a$$
$$t_a = t_b + t_c$$

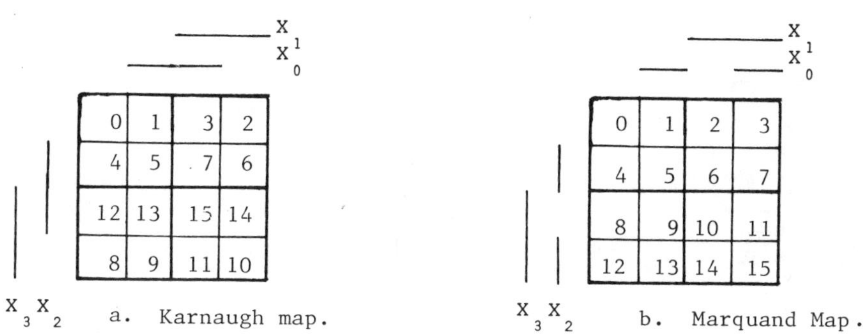

a. Karnaugh map.   b. Marquand Map.

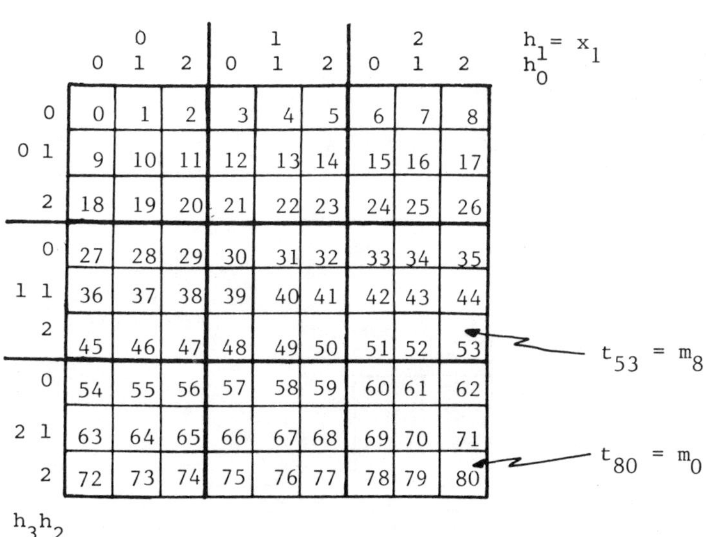

c. Triadic map.

Figure 9.3. Maps and their identifiers.

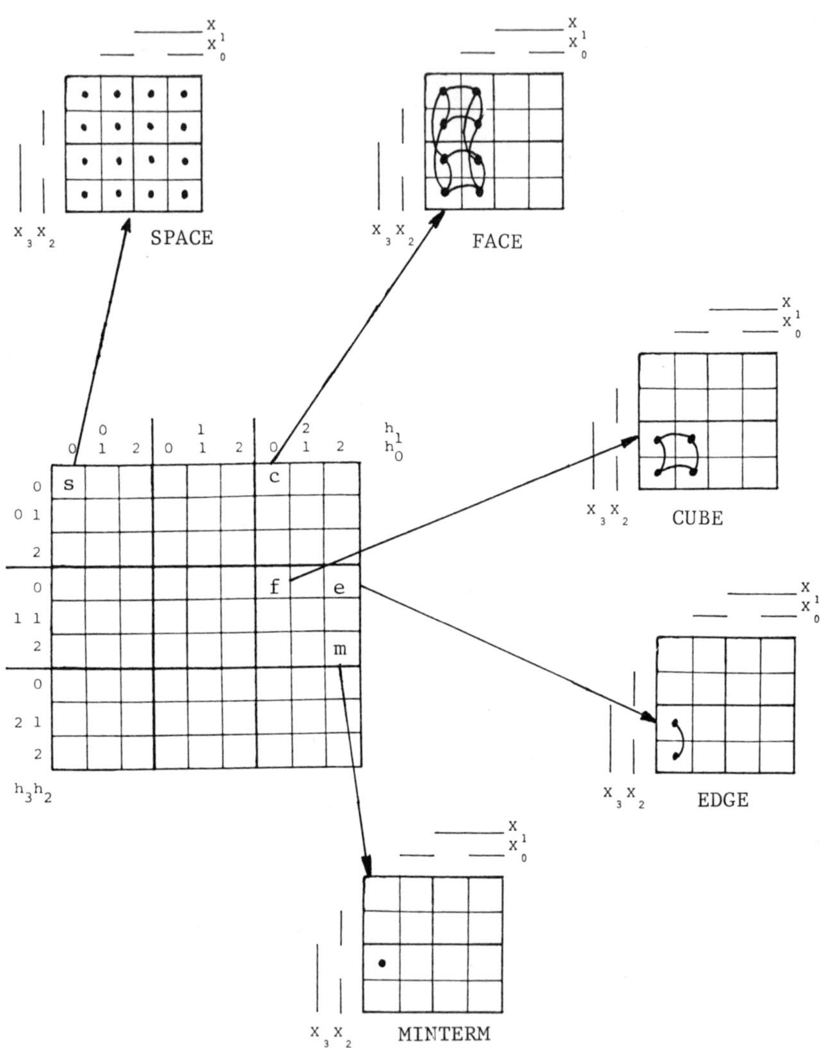

Figure 9.4.   Structures in binary and triadic space.

And also, note that

$$c - b = b - a = 3^j$$

Three points which are a triad form a triadic line in
the triadic space, as shown in Figure 9.5. Several
triads are shown in Figure 9.6. This corresponds to
the edge formed in binary space by two points whose
logical distance is 1. If t_b and t_c are minterms, then
t_a is the term representing the edge in binary space.
This and other relationships are detailed in Figure 9.7
for a four-variable space. (The four-variable space was
chosen for convenience and no limitation on problem size
is to be inferred.)

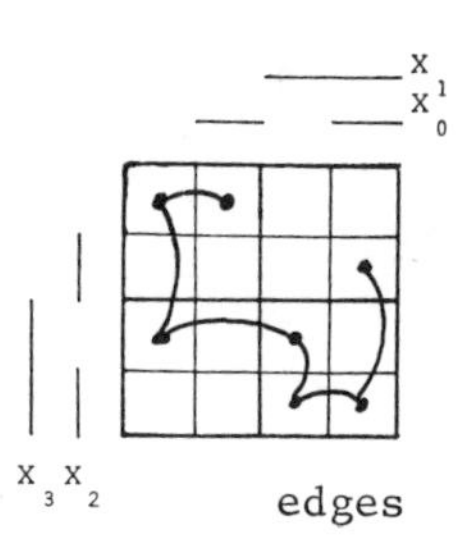

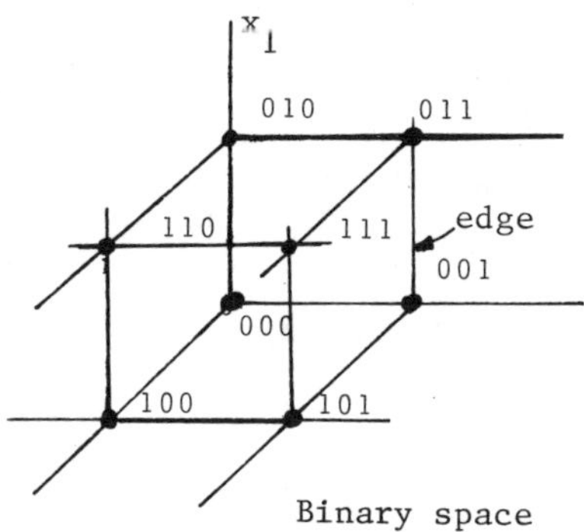

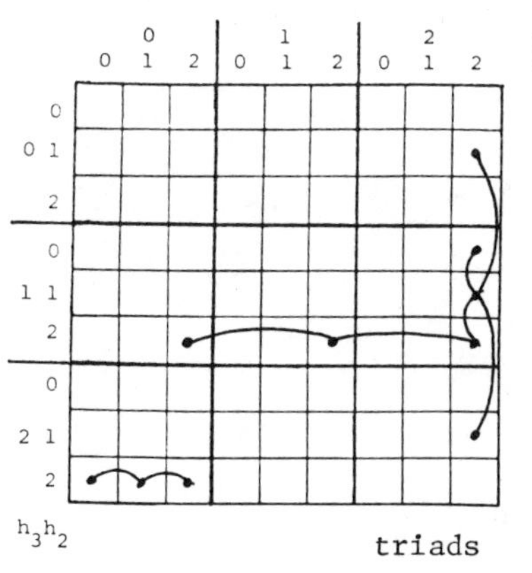

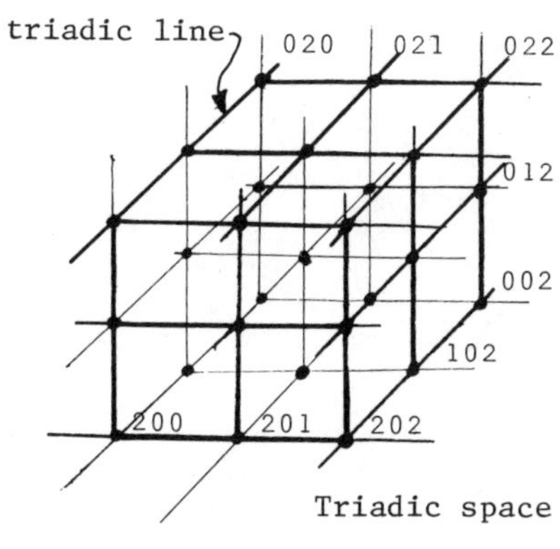

Figure 9.5. Spatial relationships.

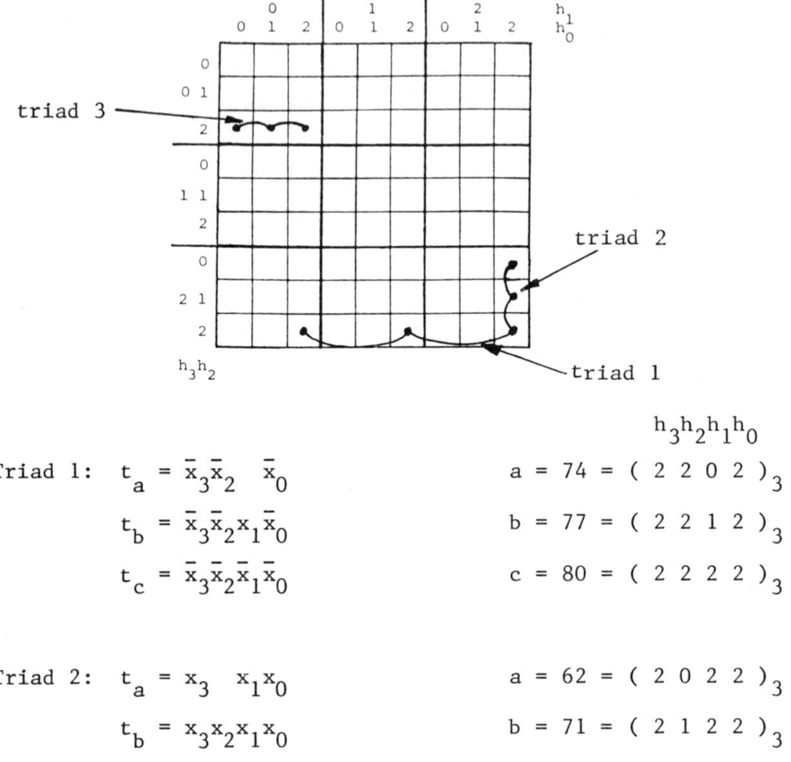

$$h_3 h_2 h_1 h_0$$

Triad 1: $\quad t_a = \bar{x}_3 \bar{x}_2 \;\; \bar{x}_0 \qquad\qquad a = 74 = (\; 2 \; 2 \; 0 \; 2 \;)_3$

$\qquad\qquad t_b = \bar{x}_3 \bar{x}_2 x_1 \bar{x}_0 \qquad\qquad b = 77 = (\; 2 \; 2 \; 1 \; 2 \;)_3$

$\qquad\qquad t_c = \bar{x}_3 \bar{x}_2 \bar{x}_1 \bar{x}_0 \qquad\qquad c = 80 = (\; 2 \; 2 \; 2 \; 2 \;)_3$

Triad 2: $\quad t_a = x_3 \;\; x_1 x_0 \qquad\qquad a = 62 = (\; 2 \; 0 \; 2 \; 2 \;)_3$

$\qquad\qquad t_b = x_3 x_2 x_1 x_0 \qquad\qquad b = 71 = (\; 2 \; 1 \; 2 \; 2 \;)_3$

$\qquad\qquad t_c = x_3 \bar{x}_2 x_1 x_0 \qquad\qquad c = 80 = (\; 2 \; 2 \; 2 \; 2 \;)_3$

Triad 3: $\quad t_a = \;\; x_2 \qquad\qquad\qquad a = 18 = (\; 0 \; 2 \; 0 \; 0 \;)_3$

$\qquad\qquad t_b = \;\; x_2 \;\; x_0 \qquad\qquad\quad b = 19 = (\; 0 \; 2 \; 0 \; 1 \;)_3$

$\qquad\qquad t_c = \;\; x_2 \;\; \bar{x}_0 \qquad\qquad\quad c = 20 = (\; 0 \; 2 \; 0 \; 2 \;)_3$

Figure 9.6.   Example triads.

9.6 INFORMATIONAL CONTENT OF MAPS

The amount of information that can be presented in each of the two graphic mediums, binary maps (Marquand, Karnaugh, Veitch) and ternary or triadic maps, highlights their differences. There are three symbols which are used for binary maps, as shown in Table 9.1; there are eight symbols for the triadic map (Table 9.2). The position of a square within a triadic map also imparts data (this is due to the ordering of implicants).

Also, note that reduction figures (edges, faces, cubes, etc.) must be drawn or visualized for the binary maps, while for the triadic map such structures are represented by a single term.

Referring to Figure 9.7, given the t_b and t_c terms of a triad, the notation for the t_a term is found using Figure 9.8. For example, if two minterms of a function form an edge, the t_a term of their triad is noted as I, an implicant of the function. If only one of the two minterms belongs to y and the other belongs to $Y = \bar{y}$, then the edge as a structure does not exist as noted by the symbol $\emptyset$.

9.7 BOOLEAN NOTATIONAL FORMS

There are several notational forms which can occur for Boolean functions. For the function $y = F(x)$, these are (1) the canonical $\Sigma\Pi$ form, (2) the $\Sigma\Pi$ form, (3) the maximal $\Sigma\Pi$ form, and (4) the complete sum. Equivalent notational forms exist for the complement function $Y = \bar{y}$.

Symbol	Definition	Name
·	$m_a \Rightarrow y$	Minterm of y
0	$m_a \Rightarrow Y$	Minterm of Y
#	$(m_a \Rightarrow y) \vee (m_a \Rightarrow Y)$	Don't care

Table 9.1. Binary map symbols.

Symbol	Definition	Name
#	$(t_h = m_a) \not\vee (t_h = m_a)$ $\wedge (t_h \rightarrow y) \not\vee (t_h \rightarrow Y)$	Don't care
·	$(t_h = m_a) \wedge (t_h \rightarrow y);$ $\sim (t_h \rightarrow Y)$	Minterm of y
0	$(t_h = m_a) \wedge (t_h \rightarrow Y);$ $\sim (t_h \rightarrow y)$	Minterm of Y
I	$(t_h \rightarrow y); \sim (t_h \rightarrow Y)$	Implicant of y
N	$(t_h \rightarrow Y); \sim (t_h \rightarrow y)$	Implicant of Y
/	$\sim (t_h \rightarrow Y) \wedge (t_h \rightarrow y)$ $\not\vee \sim (t_h \rightarrow y)$	Nonimplicant of y
o	$\sim (t_h \rightarrow y) \wedge (t_h \rightarrow Y)$ $\not\vee \sim (t_h \rightarrow Y)$	Nonimplicant of Y
∅	$\sim (t_h \rightarrow y) \wedge (t_h \rightarrow Y)$	Nonimplicant of y and nonimplicant of Y

Also used:

II	Given term of $\Sigma\Pi$-form of y
N	Given term of $\Sigma\Pi$-form of Y

Table 9.2. Triadic map symbols.

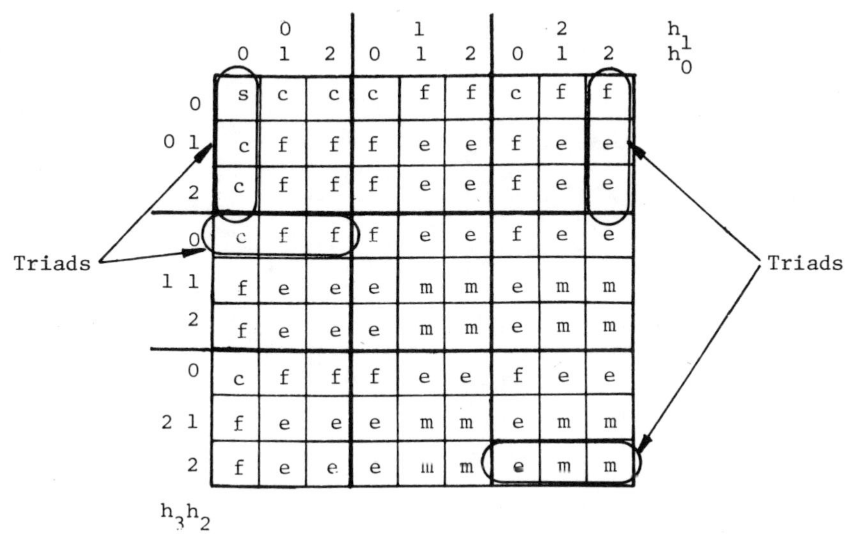

Key:

    m: minterms

    e: edge: edges are t_a terms of triads whose t_b and t_c terms are minterms.

    f: face: faces are t_a terms of triads whose t_b and t_c terms are edges.

    c: cube: cubes are t_a terms of triads whose t_b and t_c terms are faces. Cubes are extended to n dimentions.

    s: space:(universe)the t_a term of triads whose t_b and t_c terms are cubes where $n' = n - 1$.

Figure 9.7. Detail of term relationships for a four-variable map.

141

CHAPTER 9

<table>
<tr><td colspan="2"></td><td colspan="8">t_c ⟶</td></tr>
<tr><td colspan="2"></td><td></td><td colspan="2">II</td><td colspan="2">N</td><td></td><td></td><td></td></tr>
<tr><td colspan="2">t_a</td><td>#</td><td>•</td><td>I</td><td>0</td><td>N</td><td>/</td><td>o</td><td>∅</td></tr>
<tr><td colspan="2">#</td><td>#</td><td>/</td><td>/</td><td>o</td><td>o</td><td>/</td><td>o</td><td>∅</td></tr>
<tr><td rowspan="2">II</td><td>•</td><td>/</td><td>I</td><td>I</td><td>∅</td><td>∅</td><td>/</td><td>∅</td><td>∅</td></tr>
<tr><td>I</td><td>/</td><td>I</td><td>I</td><td>∅</td><td>∅</td><td>/</td><td>∅</td><td>∅</td></tr>
<tr><td rowspan="2">N</td><td>0</td><td>o</td><td>∅</td><td>∅</td><td>N</td><td>N</td><td>∅</td><td>o</td><td>∅</td></tr>
<tr><td>N</td><td>o</td><td>∅</td><td>∅</td><td>N</td><td>N</td><td>∅</td><td>o</td><td>∅</td></tr>
<tr><td colspan="2">/</td><td>/</td><td>/</td><td>/</td><td>∅</td><td>∅</td><td>/</td><td>∅</td><td>∅</td></tr>
<tr><td colspan="2">o</td><td>o</td><td>∅</td><td>∅</td><td>o</td><td>o</td><td>∅</td><td>o</td><td>∅</td></tr>
<tr><td colspan="2">∅</td><td>∅</td><td>∅</td><td>∅</td><td>∅</td><td>∅</td><td>∅</td><td>∅</td><td>∅</td></tr>
</table>

(t_b is the vertical axis, pointing downward along the rows.)

Figure 9.8. Triadic map symbol intersection chart.
(Note that some intersections are impossible but are defined.)

9.7.1. Canonical $\Sigma\Pi$-form.

The canonical $\Sigma\Pi$-form of a Boolean function y is the unique sum of all of the minterms for which for function $y = F(x)$ is true.

$$y = \sum_a m_a \cdot f_a \quad \text{for all } a \in \{ a \mid f_a = 1 \} = \{a\}_y$$

The Marquand and triadic maps of an example function are shown in Figure 9.9.

The canonical $\Sigma\Pi$-form of the complement Boolean function Y where $Y = \bar{y}$ is the unique sum of all of the minterms for which the function $y = F(x)$ is false.

$$Y = \sum_{a'} m_{a'} \cdot f_{a'} \quad \text{for all } a' \in \{ a \mid f_{a'} = 0 \} = \{a'\}_Y$$

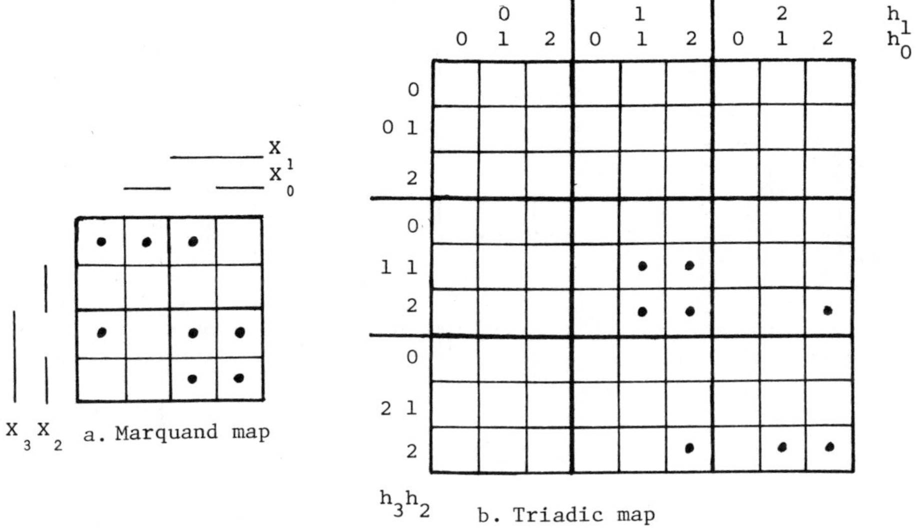

$$y = m_0 + m_1 + m_2 + m_8 + m_{10} + m_{11} + m_{14} + m_{15}$$

Figure 9.9.   Canonical $\Sigma\Pi$-form of an example function.

9.7.2 $\Sigma\Pi$-form.

The $\Sigma\Pi$-form of a Boolean function y is a not-neces-sarily-unique sum of terms for which the function is true; i.e., there may be more than one $\Sigma\Pi$-form for any given function. The sum of terms, each of which is an implicant of the function, includes sufficient terms to cover all of the minterms for which the function is true.

For y:

$$y = \sum_h t_h \text{ where every } h \in \{ h \mid t_h \rightarrow y \} \equiv \{h\}_y$$

for Y:

$$Y = \sum_{h'} t_{h'} \text{ where every } h' \in \{ h \mid t_h \rightarrow Y \} \equiv \{h\}_Y$$

The Marquand and triadic maps of the $\Sigma\Pi$-form of the pre-vious example function are shown in Figure 9.10.

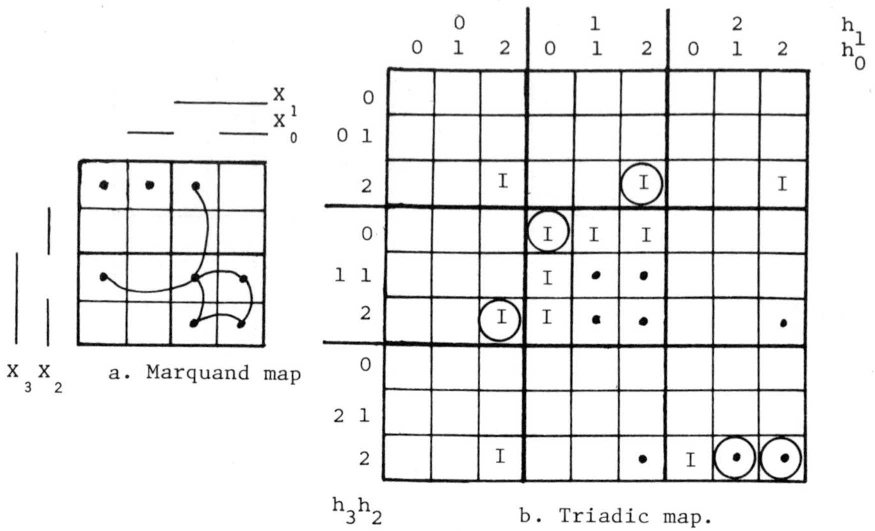

a. Marquand map

b. Triadic map.

$$y = t_{23} + t_{30} + t_{47} + t_{79} + t_{80}$$

Figure 9.10. Map of a $\Sigma\Pi$-form of y.

9.7.3. Maximal $\Sigma\Pi$-form.

The maximal $\Sigma\Pi$-form of a Boolean function y is the unique sum of all of the terms t_h for which the function y is true.

A $\Sigma\Pi$-form for y is a subset of the maximal $\Sigma\Pi$-form of y.

For y:

$$y = \sum_h t_h \quad \text{for all } h \in \{ h \mid t_h \rightarrow y \}$$
$$y = \{ t_h \}_{max}$$

For Y:

$$Y = \sum_{h'} t_{h'} \quad \text{for all } h' \in \{ h \mid t_h \rightarrow Y \}$$

The Marquand and triadic maps of the maximal $\Sigma\Pi$-form for the previous example are shown in Figure 9.11.

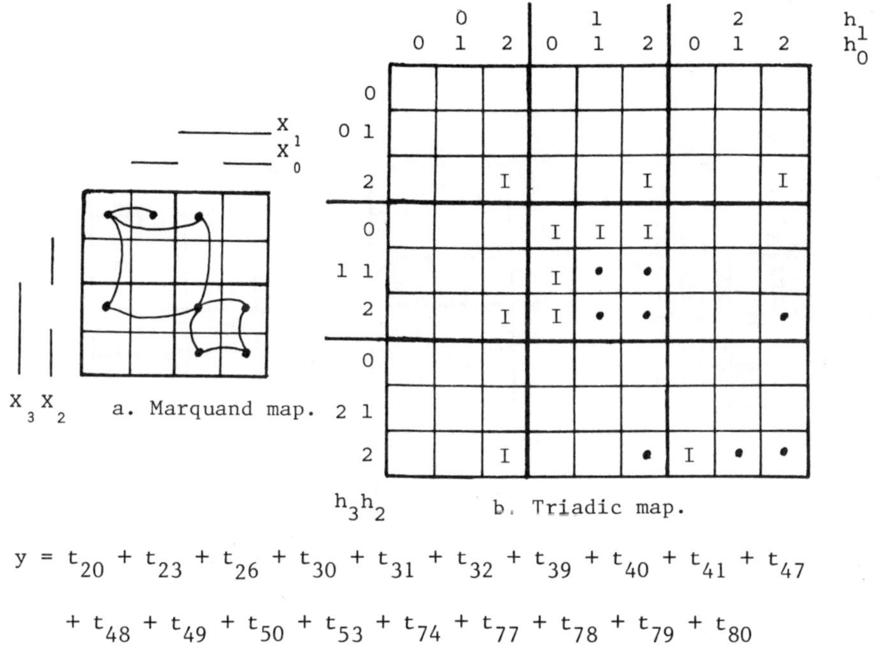

$$y = t_{20} + t_{23} + t_{26} + t_{30} + t_{31} + t_{32} + t_{39} + t_{40} + t_{41} + t_{47}$$

$$+ t_{48} + t_{49} + t_{50} + t_{53} + t_{74} + t_{77} + t_{78} + t_{79} + t_{80}$$

Figure 9.11. Map of the maximal $\Sigma\Pi$-form of y.

9.7.4. Complete sum.

The complete sum of a Boolean function y is the sum of all of the prime implicants of y:

$$y = \sum_{h} t_h \text{ where all } h \in \{ h \mid t_h \rightarrow y \} \equiv \{ h \}_y$$

$$\text{and } \sim(t_p \rightarrow t_h) \text{ for all } p \neq h$$

$$\text{where } p \in \{ h \}_y$$

$$y = \{ t_h \}_{h*}$$

The Marquand and triadic maps of the complete sum of the previous example function are shown in figure 9.12.

A prime implicant of a Boolean function y is an implicant of y which is not implied by any other implicant of the maximal $\Sigma\Pi$-form of y except itself.

There is a corresponding sum for Y.

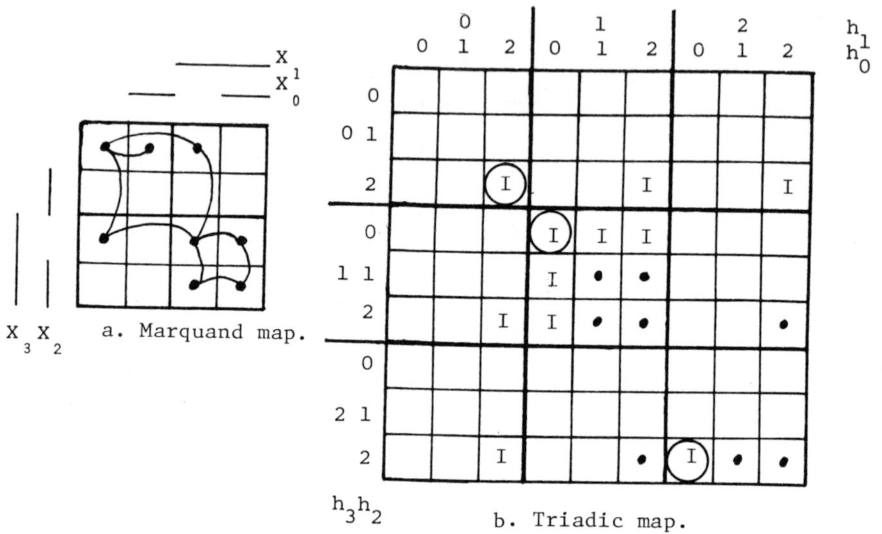

a. Marquand map.

b. Triadic map.

$$y = t_{20} + t_{30} + t_{78}$$

Figure 9.12. Map of the complete sum of y.

9.8 THE TRIADIC MAP ALGORITHM FOR FINDING THE COMPLETE SUM OF y

To solve (manually or via a computer program) a function for its complete sum, proceed as follows:

1. Map the function on the triadic map.

2. Use the triadic intersection chart (Figure 9.8) to complete the triads. It is not necessary to complete terms for Y as well as y, since the maps would be dual. The complete map for an example function is shown in Figure 9.13.

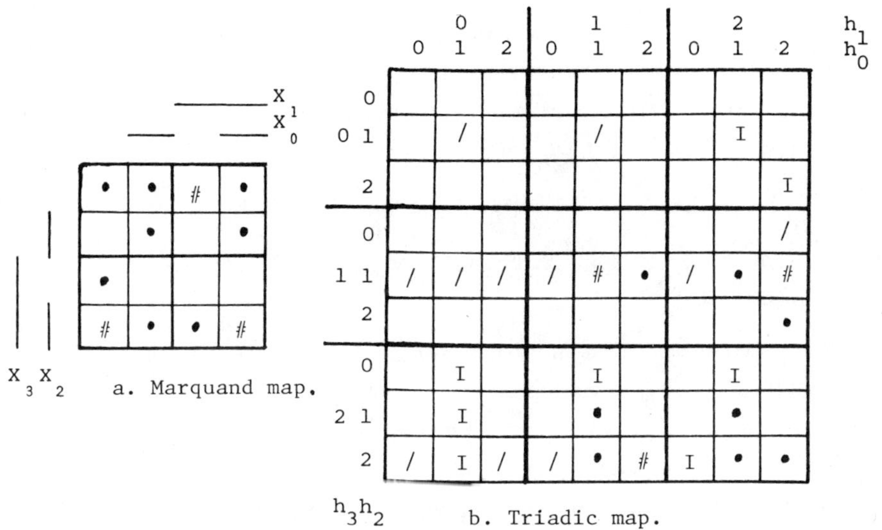

Figure 9.13.   Finding the complete sum.

3.   Find the lowest-ordered term which does not imply Y but at least _may_ imply y (denoted by / or I). This is a prime implicant. Record it and cancel it on the map. For the example function, the first term found would be $t_{10} = x_2 x_0$ (see Figure 9.14a).

4.   Cancel all t_b and t_c terms for which this term is t_a (which form a triad with the chosen term). Also, cancel terms which form a triad with any of these terms. For t_{10}, terms t_{13}, t_{16}, t_{37}, t_{41}, t_{43}, t_{64}, t_{67}, and t_{70} would be cancelled.

5. Repeat, finding the next-lowest-indexed, uncan-
celled term each time. The resulting set of
terms form the complete sum of the function y.
(See Figure 9.14b and 9.14c. The complete sum
for the example is:

$$y = t_{10} + t_{26} + t_{35} + t_{36} + t_{55} + t_{72}$$

9.9 THE PETRICK FUNCTION SOLUTION FOR THE MINIMAL $\Sigma\Pi$-FORM OF y

Once the complete sum of y has been found, by what-
ever algorithm has been chosen, the minimal $\Sigma\Pi$-form may
be found via a reduction of the prime implicant table.

The prime implicant table for the example function
of Figures 9.13 and 9.14 is shown in Figure 9.15.

The prime implicant table is a cross-reference of
the prime implicants and the minterms ($m_i \rightarrow y$) of the
function y.

From the columns of the table, one per minterm, form
the Petrick function as follows:

1. Each column contains a $\checkmark$ for each prime implicant which covers the minterm of that column.
In the example, minterm m_0 is shown to be covered
by prime implicants p_1 and p_5.

Form a sum term for each minterm using the symbol
p_i for each of the prime implicants for that minterm. In the example, the sum term for m_0 is
($p_1 + p_5$); for m_3, it is ($p_4 + p_5$). The complete
expression is shown in Figure 9.15.

2. Solve the completed equation in $\Pi\Sigma$-form by expanding it to $\Sigma\Pi$-form and selecting the term with
the minimum number of literals. This product
term has as its literals the symbols of the prime
implicants of the minimal $\Sigma\Pi$-form of y.

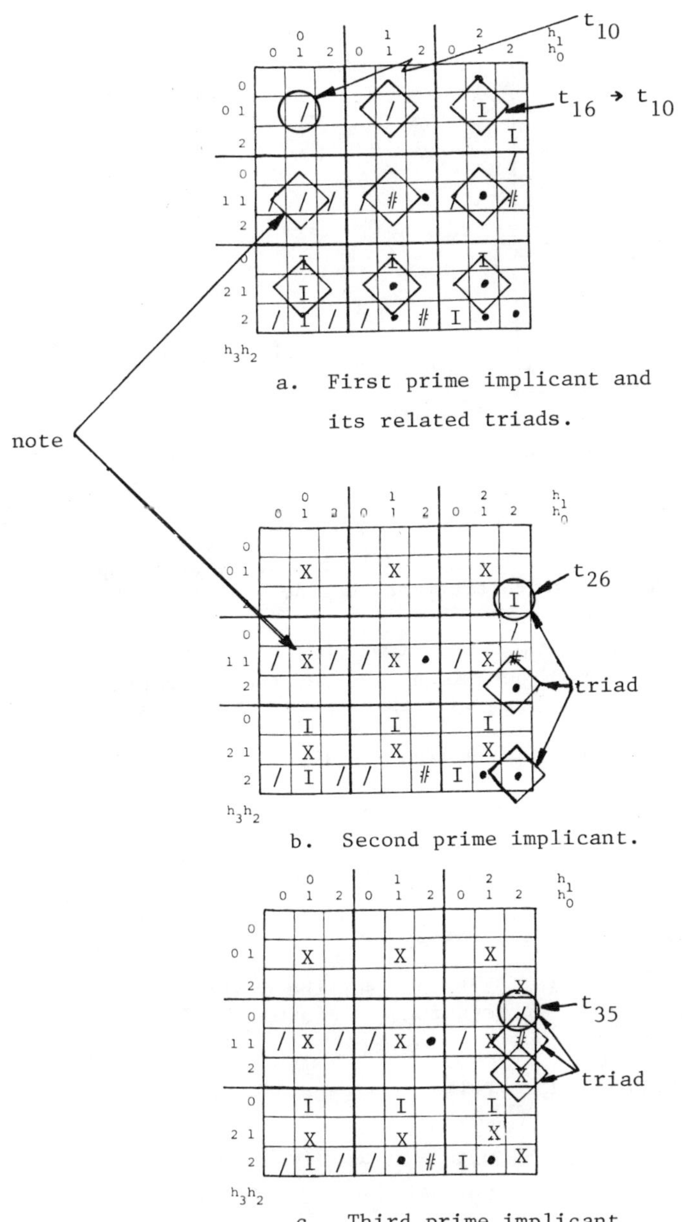

Figure 9.14. Finding prime implicants via the triadic map.

term	literals	symbol	m_0	m_1	m_3	m_5	m_7	m_8	m_{13}	m_{15}
t_{10}	x_2x_0	p_0				✓	✓		✓	
t_{26}	$\overline{x}_2\overline{x}_1\overline{x}_0$	p_1	✓					✓		
t_{35}	$x_3x_1x_0$	p_2						✓		
t_{36}	x_3x_2	p_3							✓	✓
t_{55}	$\overline{x}_3x_0$	p_4		✓	✓	✓	✓			
t_{72}	$\overline{x}_3\overline{x}_2$	p_5	✓	✓	✓					

Notes:

1. Ordering of terms, and therefore of subscripts of p_i, is arbitrary and has no effect on the solution.

2. To "cover" m_0 there is a choice of p_1 or p_5. To cover m_{15} there is no choice, therefore prime implicant x_3x_2 (p_3) is present in every solution.

$$S = (p_1 + p_5)(p_4 + p_5)(p_4 + p_5)(p_0 + p_4)(p_0 + p_4)(p_1 + p_2)(p_0 + p_3)(p_3)$$

$$\quad m_0 \qquad m_1 \qquad m_3 \qquad m_5 \qquad m_7 \qquad m_8 \qquad m_{13} \qquad m_{15}$$

Expand S into:

$$S = p_1p_3p_4 + p_1p_2p_3p_4 + \ldots$$

where $p_1p_3p_4$ is (in this case) the minimal literal product term representing the minimal $\Sigma\Pi$-form of y.

Figure 9.15. Prime implicant table.

For the example, $p_1p_3p_4$ has the least number of literals. Therefore, the minimal $\Sigma\Pi$-form of y is

$$y = \overline{x}_2\overline{x}_1\overline{x}_0 + x_3x_2 + \overline{x}_3x_0 .$$

The Boolean equation expansion is also available on the parallel Boolean processor.

9.10   LOGICAL INSTRUMENTS -- PRIME IMPLICANT GENERATION

9.10.1   Introduction

The triadic deck set is a pair of 80-column card decks which implement the triadic approach to minimization presented in Section 9.8. The first deck is for the representation of the function by its minterms, and is referred to as the binary or minterm deck. The second deck, known as the triadic or term deck, is for the representation of the function in triadic form.

The use of 80-column cards limits the function to a maximum of four variables where the term $t_0 = 1$ is not represented.

9.10.2   The Minterm Deck

The minterm deck consists of 16 cards labeled 0 through 15, one for each point identifier on a four-variable map (one for each point in a four-variable binary space).

Each card has a punch in every column in either row 0 or row 7. A 7-punch is in any column whose numeric label (1 through 80) is the same as the term identifier for a term implied by the minterm.

$$(\text{7-punch in column a of card i}) \equiv (m_i \rightarrow t_a)$$

A 0-punch is used in every column which represents a term <u>not</u> implied by the minterm.

$$(\text{0-punch in column a of card i}) \equiv \sim(m_i \rightarrow t_a)$$

A 7-punch is referred to a non-punch with reference to row 0.

A triadic map is used to explain the cards throughout this section. Figure 9.16 is a map representing the punch-nonpunch configuration for card number 4 (which represents minterm m_4).

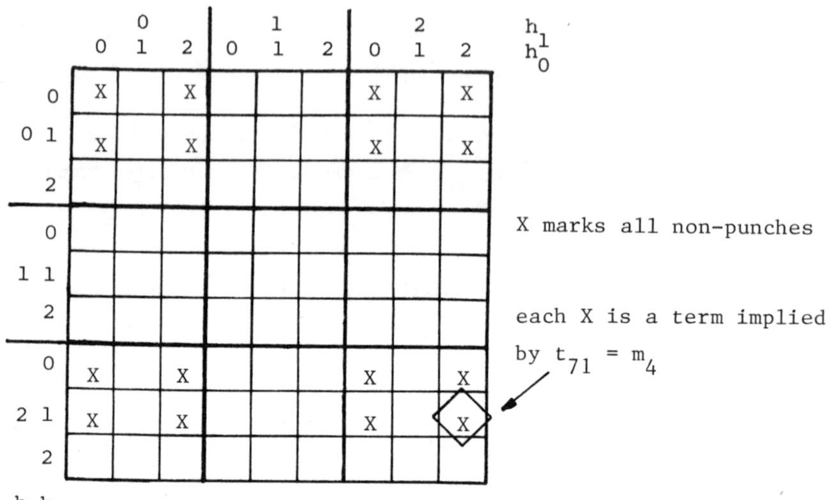

Figure 9.16. Map of minterm card 4.

The Marquand map of an example function is shown in Figure 9.17. To use the deck, proceed as follows:

1. Rather than choose all cards from the minterm deck where $p_i = 1$ or $p_i = \#$, choose all cards where $p_i = 0$. The minterms chosen are the minterms of Y where $Y = \bar{y}$. For the example function, cards 3, 4, 5, 6, 7, 9, 12, and 13 would be chosen.

2. Form a deck with these cards. Hold the deck to a light with the cards face-up and with the 9-edge down.

3. Observe the punches in the 0-row. All columns which have no punches visible are terms which <u>do</u> <u>not</u> belong to the maximal $\Sigma\Pi$-form of y. All "windows" in the 0-row represent terms which <u>do</u> belong to the maximal $\Sigma\Pi$-form of y. (The triadic map in Figure 9.18 is labeled with the term identifiers of these terms for the example function.)

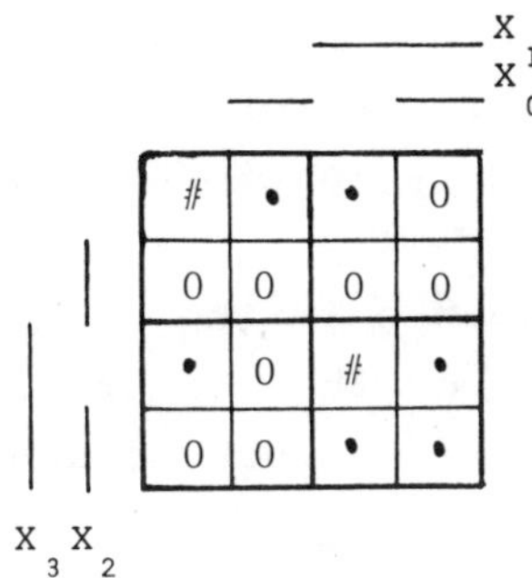

Figure 9.17. Example function.

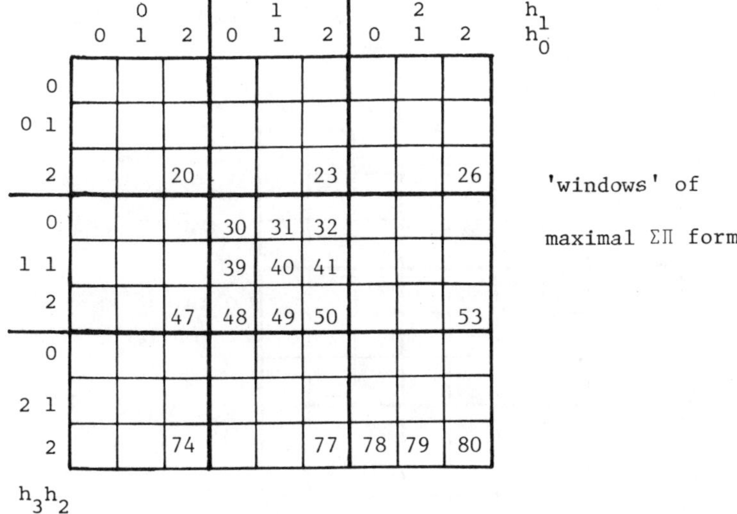

'windows' of

maximal $\Sigma\Pi$ form

Figure 9.18. The maximal $\Sigma\Pi$-form of y.

9.10.3   The Term Deck

The term deck contains 80 cards labeled 1 through 80 for each of the term identifiers for terms in a four-variable triadic space. (Term t_0 is excluded as trivial as it represents $y = 1$.)

Each card contains punches in the column corresponding to the term identifier of the term represented by the card. These are in rows 3 (index punch) and 0 (prime implicant punch).

Each card has been punched in row zero in each column that represents a non-implicant of the term which the card represents.

Each card is non-punched in row 0 in each column that represents an implicant of the term which the card represents. The exception is the prime implicant punch referenced above.

Figure 9.19 is a triadic map representing the punch/non-punch configuration for card number 10 (which represents term t_{10}).

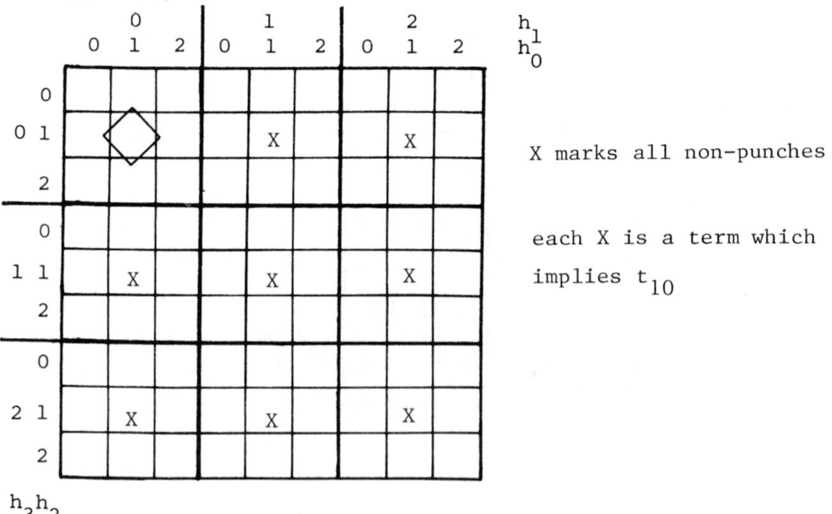

Figure 9.19. Map of triadic card 10.

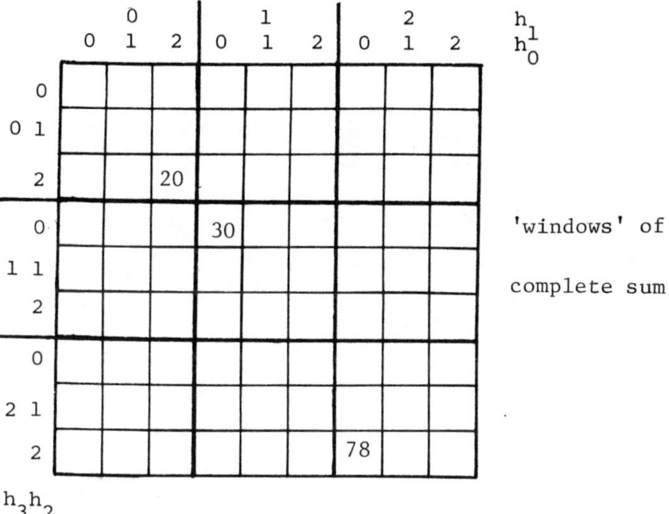

Figure 9.20.   The complete sum.

To continue the example of Section 9.10.2:

4.   Choose one triadic term card for each "window"
     or term in the set of implicants of y (one card
     for each term in the maximal $\Sigma\Pi$-form of y).

5.   Add this set of cards to the set formed from
     the minterm deck and, keeping them face-forward
     and 9-edge down, hold them to a light so that
     the holes may be examined.

6.   This time, all columns which have visible
     "windows" represent terms which are prime
     implicants of y. This is the complete sum of
     y. (The complete sum is not necessarily the
     minimal $\Sigma\Pi$-form.) The triadic map of the com-
     plete sum for the example function is shown in
     Figure 9.20.

7. To find the minimal $\Sigma\Pi$-form of y, the cards are
    used to generate the prime implicant table. Take
    from the <u>unused</u> cards of the minterm deck those
    cards which represent only the "1"s of the func-
    tion, i.e., the minterms of the function y.

8. <u>One</u> <u>at</u> <u>a</u> <u>time</u>, place these cards upside down
    behind the combined decks (i.e., 9-edge up, face
    down).

9. The remaining visible windows represent the prime
    implicants which cover the minterm represented
    by the inverted card. (The inversion uses the
    punches that are or are not present in row 7.)

10. Record this information in the prime implicant
    table, <u>remove</u> <u>the</u> <u>inverted</u> <u>card</u>, and proceed to
    the next until the table is completed.

11. Find the minimal $\Sigma\Pi$-form using the Petrick func-
    tion.

10.1   INTRODUCTION

The triadic notation introduced in Chapter 9 is used
heavily in the two previously-published papers on the
parallel Boolean processor. The triadic map (see Figure
9.3) is used here to pictorially demonstrate the basic
theorems. Detailed proofs of the theorems are not presented (they are available in the referenced papers).
The table of intersections(Table 9.2) is based on the
theorems.

10.2   THE THEOREMS

10.2.1   Theorem 3.1

Theorem 3.1 states that if the set of triadic points
represented by the set of point identifiers {a,b c} form
a triad, then if the triadic terms t_b and t_c both imply
a function y, the triadic term t_a implies y. Also, if t_a
implies y then both t_b and t_c imply y.

$$(t_a \rightarrow y) \equiv [(t_b \rightarrow y) \cap (t_c \rightarrow y)]$$

$$t_b \rightarrow t_a \rightarrow y \qquad t_c \rightarrow t_a \rightarrow y$$

The theorem also holds for the complement function $Y = \bar{y}$:

$$(t_a \rightarrow Y) \equiv [(t_b \rightarrow Y) \cap (t_c \rightarrow Y)]$$

Figure 10.1 demonstrates Theorem 3.1. The term t_{24} is the t_a of a triad formed by points $\{24,25,26\}$. Since t_{24} is a given term of the function y, it implies y and therefore both t_{25} and t_{26} imply y.

Points $\{39,40,41\}$ form a triad. t_{40} and t_{41} are given terms of the function y, and therefore t_{39} must imply y.

The notation for the complement space in terms of the function Y is given in Figure 10.2.

10.2.2   Theorem 3.2

Theorem 3.2 states that if the set of triadic points represented by the set of point identifiers $\{a,b,c\}$ form a triad, then if the triadic term t_b is a nonimplicant of (does not imply) function y, then the triadic term t_a is a nonimplicant of y. Also, if t_c is a nonimplicant of y, then t_a is a nonimplicant of y.

$$\sim(t_b \nrightarrow y) \rightarrow \sim(t_a \nrightarrow y)$$
$$\sim(t_c \nrightarrow y) \rightarrow \sim(t_a \nrightarrow y)$$

It is not necessary for both t_b and t_c to be nonimplicants of y for t_a to be a nonimplicant of y.

The theorem also applies to the complement space:

$$\sim(t_b \nrightarrow Y) \rightarrow \sim(t_a \nrightarrow Y)$$
$$\sim(t_c \nrightarrow Y) \rightarrow \sim(t_a \nrightarrow Y)$$

Using the notations of Figures 10.1 and 10.2, and adding to it to obtain

$\amalg$	a form term for the $\Sigma\Pi$-form of y
N	a form term for the $\Sigma\Pi$-form of Y
I	a term added using Theorem 3.1 from the given terms shown as $\amalg$
N	a term added using Theorem 3.1 from the given terms shown as N

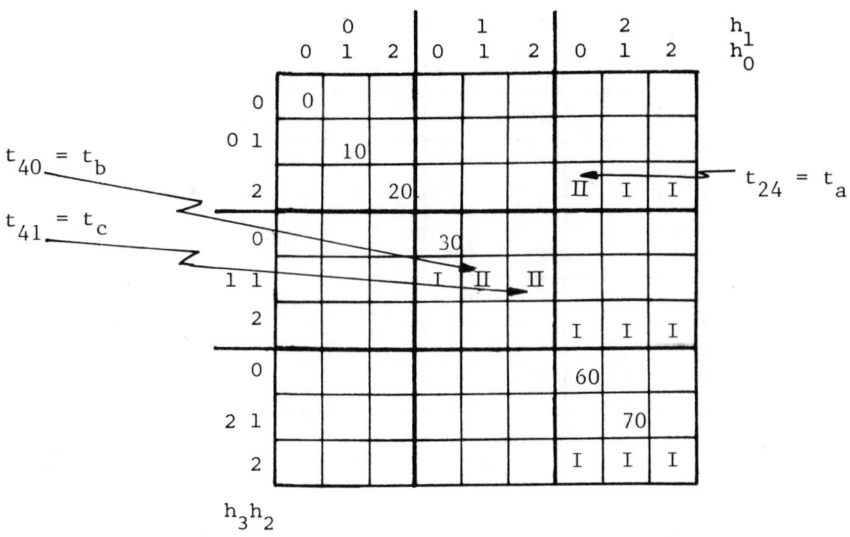

$$y = \bar{x}_2\bar{x}_1 + x_3x_2x_1x_0 + x_3x_2x_1\bar{x}_0 = t_{24} + t_{40} + t_{41}$$

t_{24}: $(t_a => y)$ therefore $(t_b => y) \cap (t_c => y)$

$t_{40}; t_{41}$: $(t_b => y) \cap (t_c => y)$ therefore $(t_a => y)$

Figure 10.1. Theorem 3.1.

$h_3 h_2$ \ $h_1 h_0$		0			1			2		
		0	1	2	0	1	2	0	1	2
0	0	0								
	1		10							N
	2			20	N	N	N			
1	0									
	1									N
	2				N	N	N			
2	0									
	1								70	N
	2				N	N	N			80

$$Y = x_2 \bar{x}_1 \bar{x}_0 + \bar{x}_2 x_1$$

$$= t_{17} + t_{21} \qquad N$$

Additional terms: N

Figure 10.2. Theorem 3.1 in complement space.

0   a term added using Theorem 3.2 where the term is a nonimplicant of y

(blank)   a blank is a "don't care" term

/   a term added using Theorem 3.2 where the term is a nonimplicant of Y

$\emptyset$ a term which is a nonimplicant of y and a nonimplicant of Y

Figure 10.3 presents a triadic map which combines Figures 10.1 and 10.2 and adds the notations derived by applying Theorem 3.2 to the expressions for y and for Y. A Marquand map is also given for reference. This is an incompletely specified function in that there are points which are considered "don't care".

10.2.3 Theorem 3.3

Theorem 3.3 states that given the triadic term t_h, which is an implicant of the function Y, and given any other triadic term, if the sum of the coefficients of the identifiers forms non-3 in all positions, then the second term is a nonimplicant of y. If the sum for any coefficient position is 3, no conclusion is made about the second term.

$$\text{if } t_h \in \{t_h\}_Y$$
$$h = (h_{n-1}\, h_{n-2} \cdots h_1\, h_0)_3$$

where h_i is a coefficient of h in the ith position, then

$$(h_j + h_j^* \neq 3 \text{ for every } j) \rightarrow [(t_h \rightarrow Y) \rightarrow \sim(t_h^* \rightarrow y)]$$

This theorem forms the basis for the hardware design of the parallel Boolean processor. It is demonstrated with a triadic map in Figure 10.4.

$$y = \bar{x}_2\bar{x}_1 + x_3x_2x_1x_0 + x_3x_2x_1\bar{x}_0$$

$$Y = x_2\bar{x}_1\bar{x}_0 + \bar{x}_2\bar{x}_1$$

$$t_{62} \not\Rightarrow y \quad \cap \quad t_{62} \not\Rightarrow Y$$

$$t_{71} \not\Rightarrow y \quad \cap \quad t_{71} \Rightarrow Y$$

$$t_{80} \Rightarrow y \quad \therefore \quad t_{80} \not\Rightarrow Y$$

Figure 10.3. Theorem 3.2.

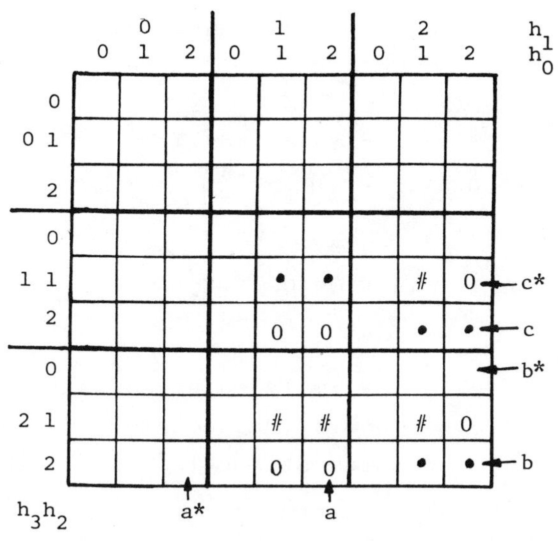

$$\begin{array}{c|cccccccc}
 & h_3 h_2 h_1 h_0 \\
\end{array}$$

	h_3	h_2	h_1	h_0	
a*	2	2	0	2	$t_a \Longrightarrow Y \therefore t_a* \not\Longrightarrow y$
a	2	2	1	2	
	4	4	1	4	h_j = NON3
b*	2	0	2	2	$t_b \Longrightarrow y \therefore t_b* \not\Longrightarrow Y$
b	2	2	2	2	
	4	2	4	4	h_j = NON3
c*	1	1	2	2	$\therefore$ no conclusion about t_c*
c	1	2	2	2	
	1	3	4	4	h_j = 3 = h_2
a*	2	2	0	2	$t_b \Longrightarrow y \quad \therefore \quad t_a* \Longrightarrow Y$
b	2	2	2	2	
	4	4	2	4	h_j = NON3

Figure 10.4. Theorem 3.3.

10.2.4 Theorem 3.4

The "ordering of implicants" referred to earlier is stated in Theorem 3.4:

For any two triadic terms t_h and t_h^*, if t_h implies t_h^* then the point identifier (triadic index) h is not smaller than the point identifier h*.

$$(t_h \rightarrow t_h^*) \rightarrow (h \geq h^*)$$

This is obvious from an examination of Figures 9.4 and 9.7.

10.2.5 Theorem 3.5

Theorem 3.5 clarifies prime implicants and is the basis of the prime implicant generation technique demonstrated in Chapter 9. The theorem states that any term in the maximal $\Sigma\Pi$-form of y which does not imply any other term in the maximal $\Sigma\Pi$-form whose triadic index is less than its own is a prime implicant of the function (see Figure 9.14).

$$\text{given } \{t_h\}_{max} \text{ is the set of terms of the maximal}$$
$$\Sigma\Pi\text{-form of } y,$$
$$t_h \in \{t_h\}_{max} \text{ and } t_p \in \{t_h\}_{max}$$
$$[\sim(t_p \rightarrow t_h) \text{ for every } t_h \text{ where } h < p]$$
$$\rightarrow (t_p \text{ is a prime implicant of } y)$$

10.2.6 Theorem 3.6

Theorem 3.6 further defines prime implicants: Given that the set $\{t_h\}_{h*}$ represents terms from the complete sum of y (the set of all prime implicants of y) such that

$$h < h^*;$$

$$h = (h_{n-1}\, h_{n-2}\, h_{n-3} \cdots h_1\, h_0)_3,$$

$$h^* = (h^*_{n-1}\, h^*_{n-2}\, h^*_{n-3} \cdots h^*_1\, h^*_0)_3,$$

then if at least one t_h from this set exists for which:

$$(h_j + h^*_j \ne 3)$$

and $(h^*_j \ne 0) \cup (h = 0)$ for every j,

then the term t^*_h is not a prime implicant of y.

10.2.7 Theorem 3.7

The last two theorems, 3.7 and 3.7A, are used in the hardware algorithms. For two triadic terms t_h and t_h^*, if the sum of the coefficients of their identifiers is not equal to 3 in all positions while the corresponding coefficient position of h* is not equal to 0, then t_h implies t_h^*. (Figure 10.5)

For two terms t_h, t_h^*,

$$\{[(h_j + h^*_j \ne 3) \cap (h^*_j > 0)]\ \text{for every } j\}$$

$$\rightarrow (t_{h^*} \rightarrow t_h)$$

10.2.8 Theorem 3.7A

This theorem modifies Theorem 3.7. Given $\{t_h\}_{\max Y}$, the set of prime implicants for the function $Y = \bar{y}$, then for any term t_h from this set and any term t_h*, if the sum of the coefficients of h and h* is not equal to 3 in all positions while the corresponding coefficient of h* is not equal to 0, then t_h* is a nonimplicant of y.

$$\{[(h_j + h^*_j \ne 3) \cap (h^*_j > 0)]\ \text{for every } j\}$$

$$\rightarrow \sim(t_h* \rightarrow y)$$

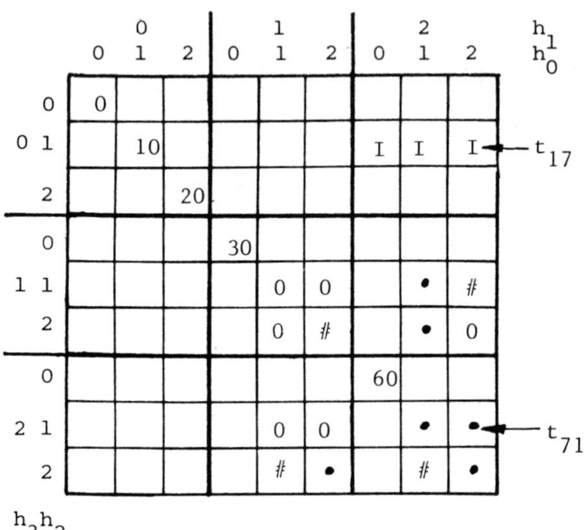

$$
\begin{array}{c}
t_{71} \\
t_{17}
\end{array}
\quad
\begin{array}{cccc}
2 & 1 & 2 & 2 \\
0 & 1 & 2 & 2 \\
\hline
2 & 2 & 4 & 4
\end{array}
\quad t_{71} \implies t_{17}
$$

$$
\begin{array}{c}
t_{17} \\
t_{15}
\end{array}
\quad
\begin{array}{cccc}
0 & 1 & 2 & 2 \\
0 & 1 & 2 & 0 \\
\hline
0 & 2 & 4 & 2
\end{array}
\quad t_{17} \implies t_{15}
$$

Figure 10.5.   Theorem 3.7.

10.3 THE ORIGINAL DESIGN OF THE PARALLEL BOOLEAN PROCESSOR

10.3.1 Introduction

The preceeding theorems and the previously-presented material on triadic maps form the basis for the parallel Boolean processor designed by Svoboda. The processor has been described in the literature under the name of the Boolean analyzer.

The parallel Boolean processor is a special processor designed to solve certain fundamental problems. Some of these are:

1. The listing of prime implicants
2. The solution of general systems of
   Boolean equations
3. The solution of special design automation
   problems
4. The generation of a test sequence for
   combinational circuits
5. The generation of the Boolean difference
   for x_i for a function F
6. The coverage problem

It was originally designed to be interfaced to the "variable" system at UCLA. Updated designs have been issued by students for MSI-level implementation. It now appears that a reasonably-sized processor could be built from LSI chips and an interface to a microprocessor system with disk memory and some form of input/output (possibly a printing terminal). With the changes in technology, the cost has become more than reasonable for such a device.

The original design was for 2 MHz operation. Present technology would allow a faster clock and therefore faster operation.

10.3.2 The Original Design

The first design called for 100 processing registers
that were to be operated in parallel. These were to be
loaded (via software interface to a main system) with the
triadic equivalent of terms from the complement space.
The processor works on the complement to the function,
with the result left in main memory as:

> 1 = cancelled
>
> 0 = implicant or prime implicant

Each register containts the storage for the coefficients
of a point in a 22-variable space stored in triadic notation. Figure 10.6 presents the original configuration
of the processor and its registers.

Each coefficient is represented by its triadic digit,
and this requires two flip-flop (F/F) elements per digit:

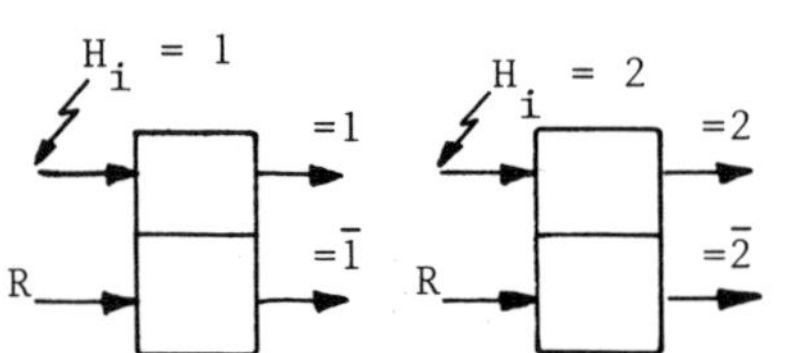

The encoding on the bus is:

$$H_i = 2$$
$$H_i = 1 \qquad \text{for the } i^{th} \text{ digit}$$
$$H_i = \bar{0}$$

This requires 44 F/F elements per register or a total of
4400 F/Fs for the register network. The details of the
processing registers are diagrammed in Figures 10.7 and
10.8.

The triadic or binary space (depending upon the
algorithm being used) is represented by storing one bit
per point in the space and by addressing this bit via a
triadic or binary "clock" which is encoded in triadic
notation. The bus from the clock contains three signal
lines per digit, with the maximum number of digits fixed
at 22.

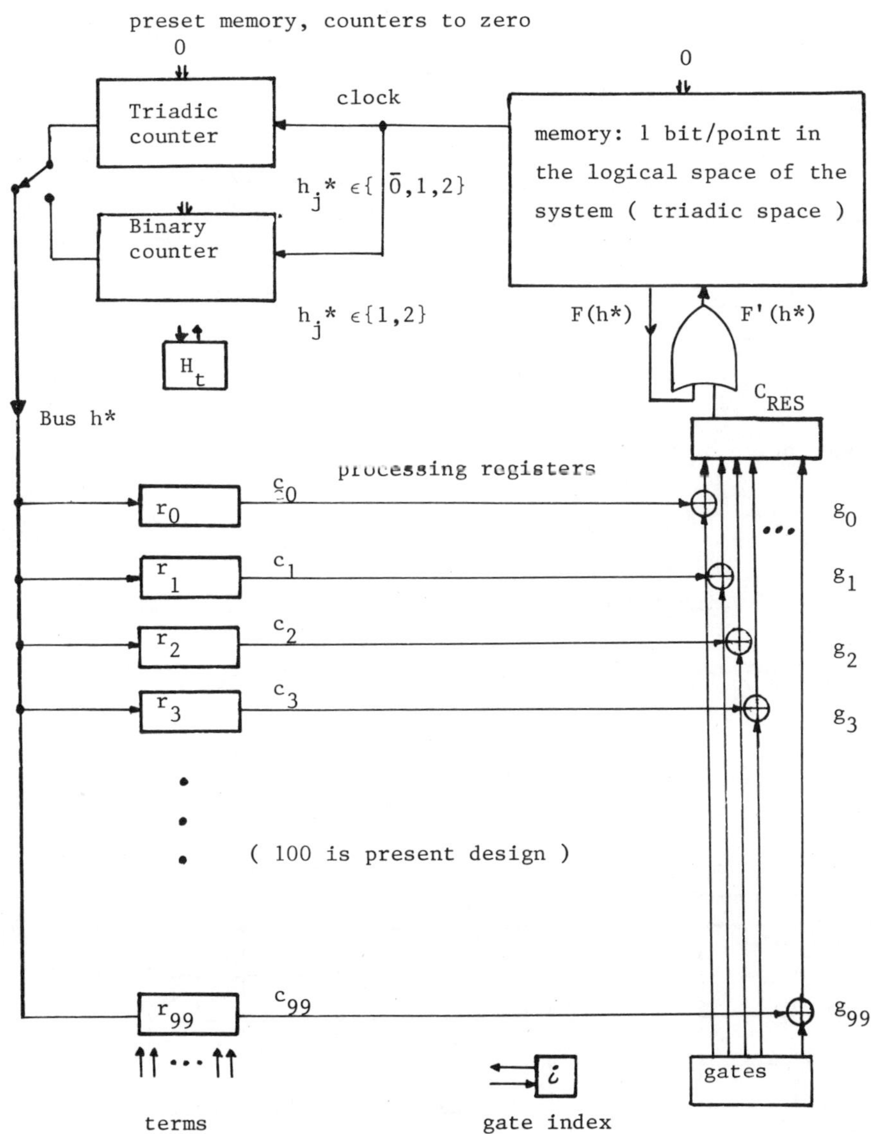

Figure 10.6. Block diagram – Parallel Boolean Processor.

Using the notation of the original paper:

$$Q = q_{prev\ digit}\ (\ (\ h_0{}^* = \bar{0}\ )\ \cup\ (\ h_0 = \bar{1}\ )(\ h_0 = \bar{2}\ )\ )$$

Implementation of theorem 3.6:

$$(\ (\ h_j{}^* = N\emptyset N0\ )\ \cup\ (\ h_j = 0\ )\ \text{for every } j\ )$$

$$\bar{P} = p_{prev\ digit}\ +\ (\ h_0 = 1\ )(\ h_0{}^* = 2\ )\ +\ (\ h_0 = 2\ )(\ h_0{}^* = 1\ )$$

Implementation of theorem 3.3:

$$(\ h_j + h_j{}^* = N\emptyset N3\ \text{for every } j\ )$$

$$c^k = p^k \quad \text{implicant listing}$$

$$c^k = p^k \cap Q^k \quad \text{prime implicant}$$

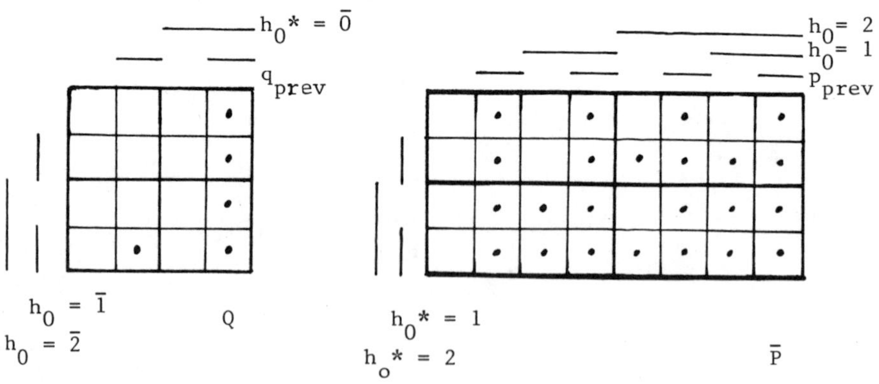

Figure 10.7. Processing register detail.

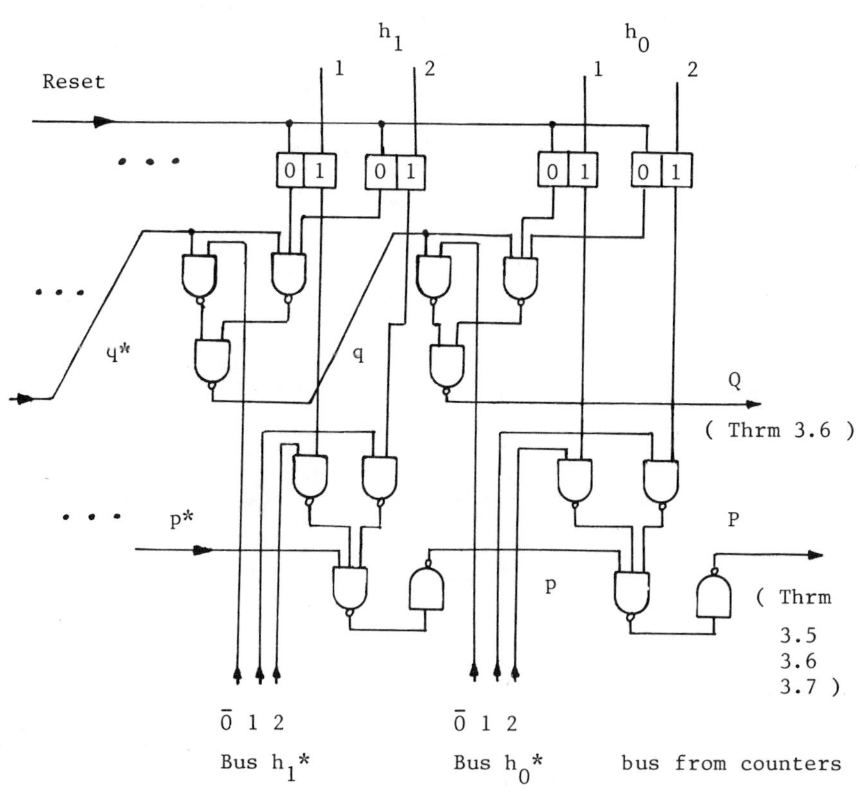

q* , p*   at unused high order positions are set ON

Figure 10.8.   Original design of the two lower
digits for the processing registers.

The original design provides for the comparison of the digits of the bus with each corresponding digit of each of the 100 processing registers at the same time. This parallelism of 22x100 effective comparisons produces one bit of information for each clock step. The bit is "OR"d into the appropriate position in the memory.

10.3.3   Improved Parallelism in the Processor Implementation

The original design provided for up to 22x100 simultaneous effective comparisons of triadic digits per clock step. This may be improved upon with a slight modification of the processor hardware. The modification is the addition of logic to the processing registers such that for each comparison step of the clock, 3^j bits rather than one bit of the result is completed. The clock is then stepped 3^j rather than 3^0 (triadic or binary clock).

The functions $\emptyset_{q*}(q)$ are defined as functions of the values of the lower-order digits of the terms and not of the clock digits. The clock digits are represented by the added hardware.

$$q = h_{i-1}\, 3^{i-1} + h_{i-2}\, 3^{i-2} + \ldots + h_0\, 3^0$$

$$\emptyset_{q*}(q) = \emptyset_{q*} =$$

$$1 \text{ for } h_j + h_{j*} \neq 3$$

$$0 \text{ for } h_j + h_{j*} = 3$$

$$\text{for every } j = 0,\, 1,\, \ldots,\, i-1$$

For the case 3^2, Figure 10.9 provides the map and the functions $\emptyset_{q*}$. Nine bits are produced for the result for each step of the clock. Figure 10.10 gives the actual implementation of these functions for one processing register. Each of the 100 registers would be the same.

Figures 10.11 and 10.12 present the map and functions for a 27-bit result per clock. The hardware tradeoffs make this version impractical at this time. If implemented it would mean 27×100×22 effective comparisons per clock step.

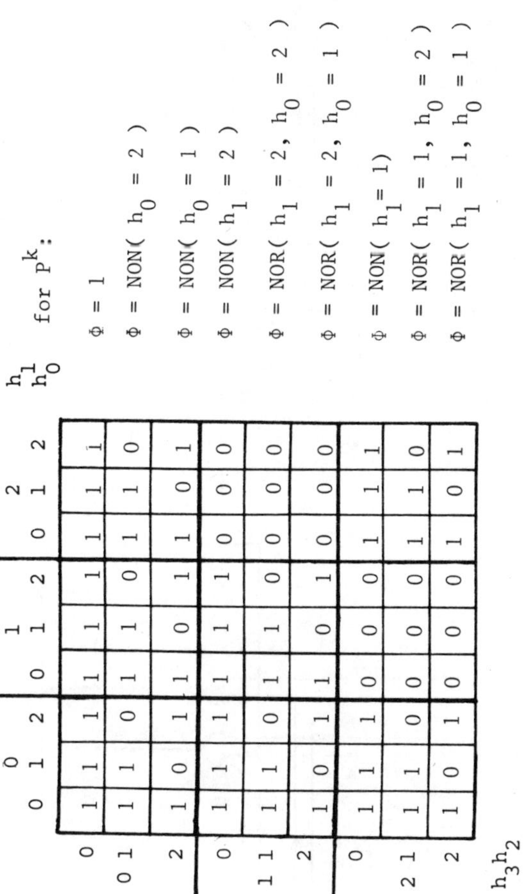

Key — h_1, h_0 for P^k:

- $\Phi = 1$
- $\Phi = \mathrm{NON}(\ h_0 = 2\ )$
- $\Phi = \mathrm{NON}(\ h_0 = 1\ )$
- $\Phi = \mathrm{NON}(\ h_1 = 2\ )$
- $\Phi = \mathrm{NOR}(\ h_1 = 2,\ h_0 = 2\ )$
- $\Phi = \mathrm{NOR}(\ h_1 = 2,\ h_0 = 1\ )$
- $\Phi = \mathrm{NON}(\ h_1 = 1)$
- $\Phi = \mathrm{NOR}(\ h_1 = 1,\ h_0 = 2\ )$
- $\Phi = \mathrm{NOR}(\ h_1 = 1,\ h_0 = 1\ )$

h_3h_2 \ h_1h_0	0 0	0 1	0 2	1 0	1 1	1 2	2 0	2 1	2 2
0 0	1	1	1	1	1	1	1	1	1
0 1	1	1	0	1	1	0	1	1	0
0 2	1	0	1	1	0	1	1	0	1
1 0	1	1	1	1	1	1	0	0	0
1 1	1	1	0	1	1	0	0	0	0
1 2	1	0	1	1	0	1	0	0	0
2 0	1	1	1	0	0	0	1	1	1
2 1	1	1	0	0	0	0	1	1	0
2 2	1	0	1	0	0	0	1	0	1

Figure 10.9. The functions Φ for a clock step of 3**2.

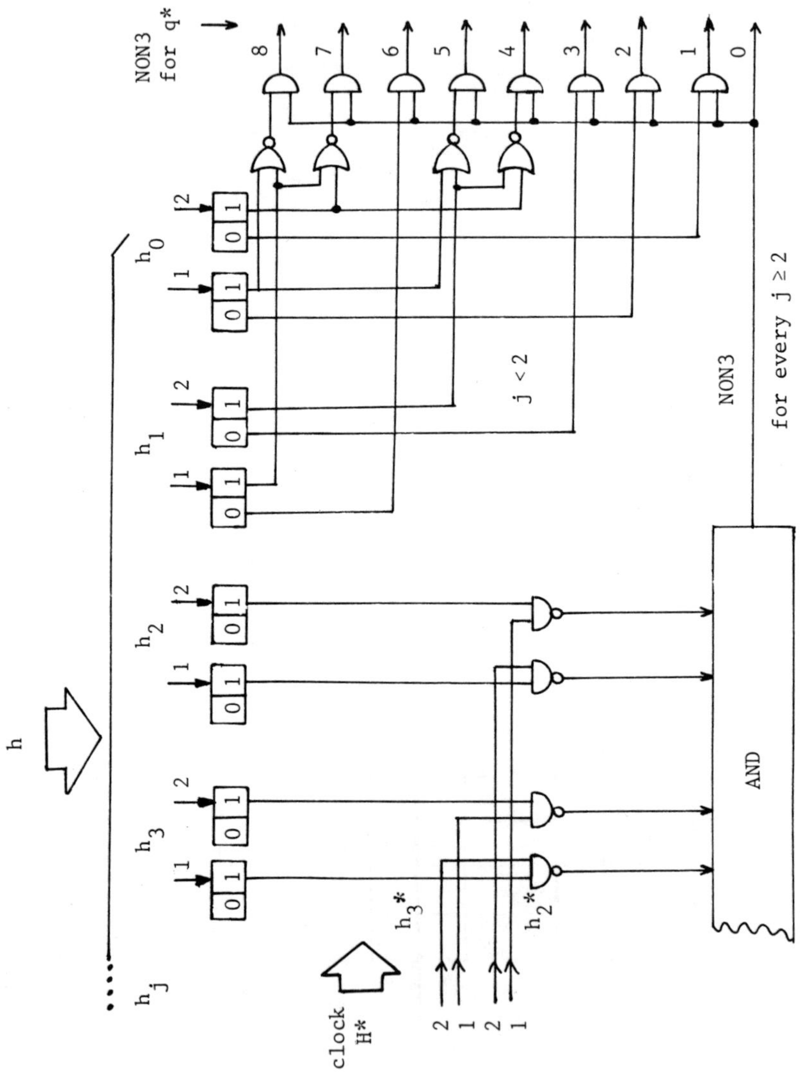

Figure 10.10.   Processing register with 9 parallel outputs.

h^*_2	h^*_1	h^*_0	0									1									2								
			0			1			2			0			1			2			0			1			2			h_1
			0	1	2	0	1	2	0	1	2	0	1	2	0	1	2	0	1	2	0	1	2	0	1	2	0	1	2	h_0
0	0	0	1	1	1	1	1	1	1	1	1	1	1	1	1	1	1	1	1	1	1	1	1	1	1	1	1	1	1
0	0	1	1	1	0	1	1	0	1	1	0	1	1	0	1	1	0	1	1	0	1	1	0	1	1	0	1	1	0
0	0	2	1	0	1	1	0	1	1	0	1	1	0	1	1	0	1	1	0	1	1	0	1	1	0	1	1	0	1
0	1	0	1	1	1	1	1	1	0	0	0	1	1	1	1	1	1	0	0	0	1	1	1	1	1	1	0	0	0
0	1	1	1	1	0	1	1	0	0	0	0	1	1	0	1	1	0	0	0	0	1	1	0	1	1	0	0	0	0
0	1	2	1	0	1	1	0	1	0	0	0	1	0	1	1	0	1	0	0	0	1	0	1	1	0	1	0	0	0
0	2	0	1	1	1	0	0	0	1	1	1	1	1	1	0	0	0	1	1	1	1	1	1	0	0	0	1	1	1
0	2	1	1	1	0	0	0	0	1	1	0	1	1	0	0	0	0	1	1	0	1	1	0	0	0	0	1	1	0
0	2	2	1	0	1	0	0	0	1	0	1	1	0	1	0	0	0	1	0	1	1	0	1	0	0	0	1	0	1
1	0	0	1	1	1	1	1	1	1	1	1	1	1	1	1	1	1	1	1	1	0	0	0	0	0	0	0	0	0
1	0	1	1	1	0	1	1	0	1	1	0	1	1	0	1	1	0	1	1	0	0	0	0	0	0	0	0	0	0
1	0	2	1	0	1	1	0	1	1	0	1	1	0	1	1	0	1	1	0	1	0	0	0	0	0	0	0	0	0
1	1	0	1	1	1	1	1	1	0	0	0	1	1	1	1	1	1	0	0	0	0	0	0	0	0	0	0	0	0
1	1	1	1	1	0	1	1	0	0	0	0	1	1	0	1	1	0	0	0	0	0	0	0	0	0	0	0	0	0
1	1	2	1	0	1	1	0	1	0	0	0	1	0	1	1	0	1	0	0	0	0	0	0	0	0	0	0	0	0
1	2	0	1	1	1	0	0	0	1	1	1	1	1	1	0	0	0	1	1	1	0	0	0	0	0	0	0	0	0
1	2	1	1	1	0	0	0	0	1	1	0	1	1	0	0	0	0	1	1	0	0	0	0	0	0	0	0	0	0
1	2	2	1	0	1	0	0	0	1	0	1	1	0	1	0	0	0	1	0	1	0	0	0	0	0	0	0	0	0
2	0	0	1	1	1	1	1	1	1	1	1	0	0	0	0	0	0	0	0	0	1	1	1	1	1	1	1	1	1
2	0	1	1	1	0	1	1	0	1	1	0	0	0	0	0	0	0	0	0	0	1	1	0	1	1	0	1	1	0
2	0	2	1	0	1	1	0	1	1	0	1	0	0	0	0	0	0	0	0	0	1	0	1	1	0	1	1	0	1
2	1	0	1	1	1	1	1	1	0	0	0	0	0	0	0	0	0	0	0	0	1	1	1	1	1	1	0	0	0
2	1	1	1	1	0	1	1	0	0	0	0	0	0	0	0	0	0	0	0	0	1	1	0	1	1	0	0	0	0
2	1	2	1	0	1	1	0	1	0	0	0	0	0	0	0	0	0	0	0	0	1	0	1	1	0	1	0	0	0
2	2	0	1	1	1	0	0	0	1	1	1	0	0	0	0	0	0	0	0	0	1	1	1	0	0	0	1	1	1
2	2	1	1	1	0	0	0	0	1	1	0	0	0	0	0	0	0	0	0	0	1	1	0	0	0	0	1	1	0
2	2	2	1	0	1	0	0	0	1	0	1	0	0	0	0	0	0	0	0	0	1	0	1	0	0	0	1	0	1

The column headers are h_2, h_1, h_0 (reading top to bottom) and the row labels are h^*_2, h^*_1, h^*_0.

Triadic map of 6 variables

$$\Phi_{q^*} = 1 \text{ iff } h_j + h_j{}^* = \text{NON3 for any } j < 3$$

$$\Phi_{q^*}(q) = \Phi_{q^*} \qquad\qquad q = h_2 3^2 + h_1 3^1 + h_0 3^0$$

Figure 10.11. Map of Φ_{q^*} where i = 3.

The functions:

each clock step produces
27 bits of output

$$\Phi_0 = 1$$
$$\Phi_1 = NON(\ h_0 = 2\ )$$
$$\Phi_2 = NON(\ h_0 = 1\ )$$
$$\Phi_3 = NON(\ h_1 = 2\ )$$
$$\Phi_4 = NOR(\ h_1 = 2,\ h_0 = 2\ )$$
$$\Phi_5 = NOR(\ h_1 = 2,\ h_0 = 1\ )$$
$$\Phi_6 = NON(\ h_1 = 1\ )$$
$$\Phi_7 = NOR(\ h_1 = 1,\ h_0 = 2\ )$$
$$\Phi_8 = NOR(\ h_1 = 1,\ h_0 = 1\ )$$
$$\Phi_9 = NON(\ h_2 = 2\ )$$
$$\Phi_{10} = NOR(\ h_2 = 2,\ h_0 = 2\ )$$
$$\Phi_{11} = NOR(\ h_2 = 2,\ h_0 = 1\ )$$
$$\Phi_{12} = NOR(\ h_2 = 2,\ h_1 = 2\ )$$
$$\Phi_{13} = NOR(\ h_2 = 2,\ h_1 = 2,\ h_0 = 2\ )$$
$$\Phi_{14} = NOR(\ h_2 = 2,\ h_1 = 2,\ h_0 = 1\ )$$
$$\Phi_{15} = NOR(\ h_2 = 2,\ h_1 = 1\ )$$
$$\Phi_{16} = NOR(\ h_2 = 2,\ h_1 = 1,\ h_0 = 2\ )$$
$$\Phi_{17} = NOR(\ h_2 = 2,\ h_1 = 1,\ h_0 = 1\ )$$
$$\Phi_{18} = NON(\ h_2 = 1\ )$$
$$\Phi_{19} = NOR(\ h_2 = 1,\ h_0 = 2\ )$$
$$\Phi_{20} = NOR(\ h_2 = 1,\ h_0 = 1\ )$$
$$\Phi_{21} = NOR(\ h_2 = 1,\ h_1 = 2\ )$$
$$\Phi_{22} = NOR(\ h_2 = 1,\ h_1 = 2,\ h_0 = 2\ )$$
$$\Phi_{23} = NOR(\ h_2 = 1,\ h_1 = 2,\ h_0 = 1\ )$$
$$\Phi_{24} = NOR(\ h_2 = 1,\ h_1 = 1\ )$$
$$\Phi_{25} = NOR(\ h_2 = 1,\ h_1 = 1,\ h_0 = 2\ )$$
$$\Phi_{26} = NOR(\ h_2 = 1,\ h_1 = 1,\ h_0 = 1\ )$$

Figure 10.12. The Φ_{q*} functions for a 3^3 clock step.

10.4 APPLICATIONS

Applications of the processor are detailed in the literature. The following sections will give a rough outline of the various procedures.

10.4.1 Implicant Listing

To find the implicants of a function, the processing registers are loaded with the terms of the complement function, $\{t_h\}_Y$ (see Figure 10.13). As the h^* counter is incremented from 0 to 3^n in steps of 3^0, representing points in the triadic space, the registers are summed with a 1 used for any NON3 sum and a 0 used for any sum containing a 3 (refer to Figure 10.14). The sum results are OR'd to form one bit of the resulting memory pattern at address h^*. The resulting bit pattern in memory is an inverse of the desired map, i.e., it contains a 0 everywhere there is a triadic term implying y. Since the results from all of the 100 registers at any clock step are OR'd, it is sufficient for 1 term to form NON3 for the term addressed in memory to be marked "cancelled".

The implicant listing application uses the triadic clock, and the triadic map.

10.4.2 Implicant Listing Under Improved Parallelism

Figure 10.15 presents the implicant listing of a function and details the register contents over the clock cycle for the case when 9 bits of results are computed per clock step. (The distinguishment of the "$\cdot$" and "I" symbols is not represented in the memory bit patterns as presented. The "$\cdot$" comes from knowing where the minterm space appears on the map.) Each clock step is 3^2 and each step produces an output equal to one row of the Marquand map in this example.

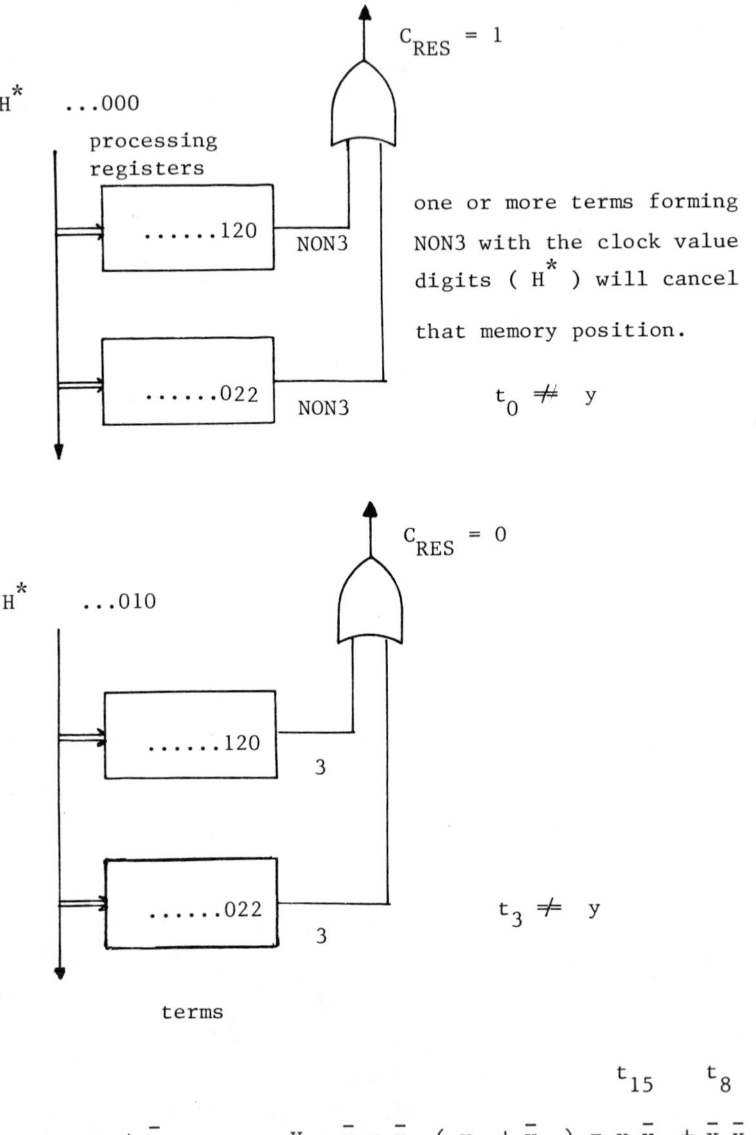

Figure 10.13.   Implicant listing – example clock step.

h* Counter	Terms	Term 1 $h^* + h = $ NON3	Term 2 $h^* + h = $ NON3	Result $F'(h^*)$
0 0 0	1 2 0	1 (1 2 0)	1 (0 2 2)	1
0 0 1	0 2 2	1	0	1
0 0 2		1	1	1
0 1 0		0 (1 3 0)	0 (0 3 2)	0
0 1 1		0	0	0
0 1 2		0	0	0
0 2 0		1	1	1
0 2 1		1	0	1
0 2 2		1	1	1
1 0 0		1	1	1
1 0 1		1	0	1
1 0 2		1 (2 2 0)	1 (1 2 4)	1
1 1 0		0	0	0
1 1 1		0	0	0
1 1 2		0	0	0
1 2 0		1	1	1
1 2 1		1	0	1
1 2 2		1	1	1
2 0 0		0	1	1
2 0 1		0	0	0
2 0 2		0	1	1
2 1 0		0	0	0
2 1 1		0	0	0
2 1 2		0	0	0
2 2 0		0	0	0
2 2 1		0	0	0
2 2 2		0 (3 4 0)	1 (2 4 4)	1

c   cancel point

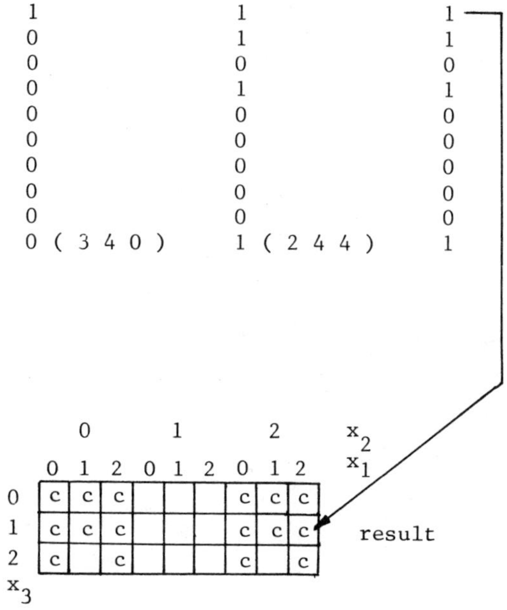

Figure 10.14.   Detail of implicant listing.

clock	Processing registers (parallel)			
h*	(2100)	(2011)	(1021)	(0120)
00--	111111111	110110000	110000110	111000111
01--	111111111	110110000	110000110	111000111
02--	000000000	110110000	110000110	000000000
10--	000000000	000000000	110000110	111000111
11--	000000000	000000000	110000110	111000111
12--	000000000	000000000	110000110	000000000
20--	111111111	110110000	000000000	111000111
21--	111111111	110110000	000000000	111000111
22--	000000000	110110000	000000000	000000000

$$\bar{y} = \bar{x}_3 x_2 + \bar{x}_3 x_1 x_0 + x_3 \bar{x}_1 x_0 + x_2 \bar{x}_1$$

Terms for the registers: (2100), (2011), (1021),

(0120)

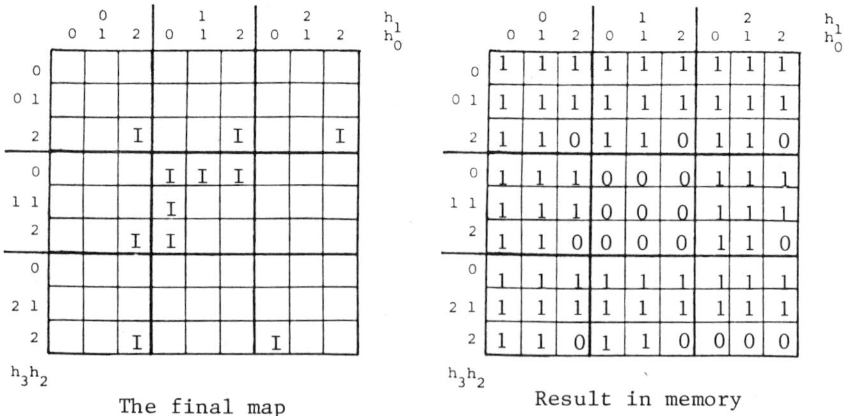

The final map   Result in memory

Figure 10.15. Implicant listing.

10.4.3   Existence Function

The Boolean processor is operated in binary mode for this application (no triadic term contains a 0). The processing is done in triadic notation.

The processing registers are loaded with the terms of Y from each of the equations of the system (100 of them) as shown in Figure 10.16. The binary counter to correspond to the memory location being addressed, the counter in triadic form, is input to bus h^*. The processing is the same as for implicant listing. A "1" is produced for any term producing a NON3 coefficient sum with the h^* bus and the results of all 100 registers are OR'd into memory.

Upon completion the complement map is stored in memory. That is, each "0" in memory is a "1" in the existence function map of the system.

The terms of Y are derived from the equations of y by the relationships:

$$(a = b) \equiv (y = 1)$$

$$(\bar{a}b + a\bar{b} = 0) \equiv (Y = 0)$$

For example, the first equation of Figure 10.16:

$$x_4 = x_0 x_1$$

produces the complement equation:

$$\overline{x_4} x_1 x_0 + x_4 \overline{x_0} + x_4 \overline{x_1} = 0$$

10.4.4   Larger Systems

10.4.4.1   More terms

Where there are more than 100 terms to be used for implicant listing, existence function generation, etc., then several passes may be made. Each pass through memory restarts the addressing bus clock h^* at the origin and is performed with a new set of 100 terms in the registers.

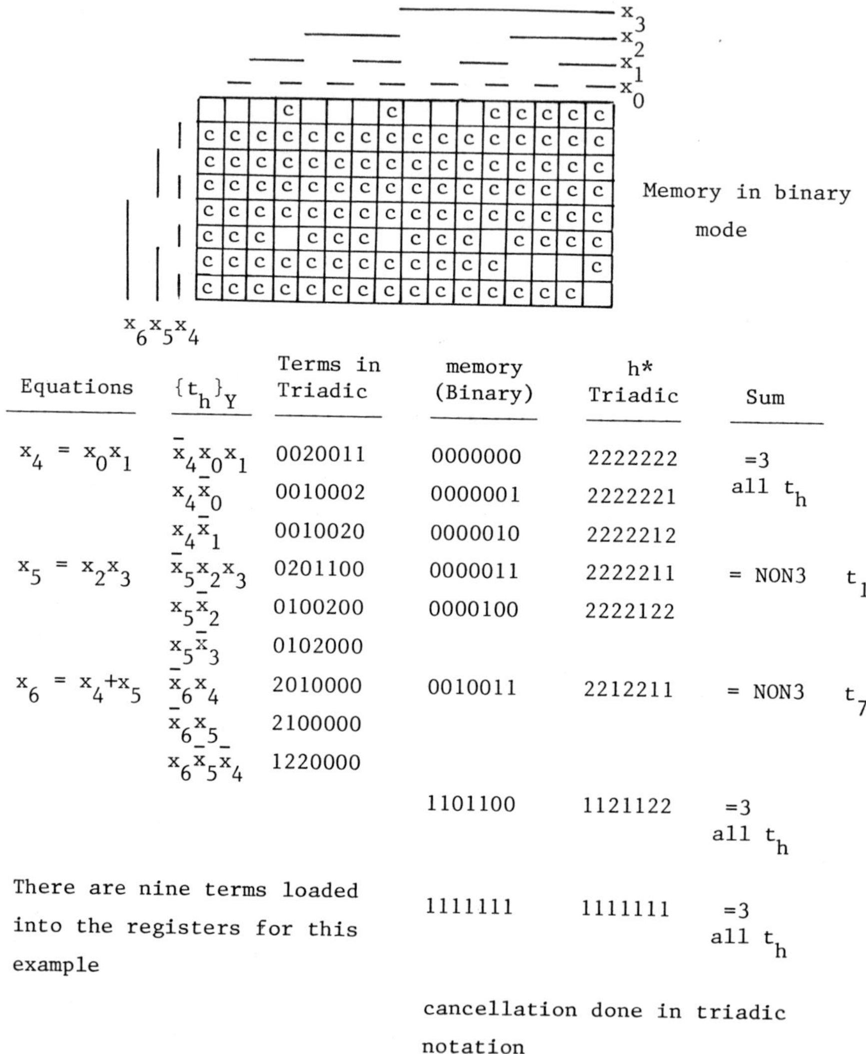

Figure 10.16. Generating the existence matrix.

As many passes as are needed may be performed until all of the terms have been fed through. There is no need to combine or reduce the Y terms of the various equations prior to processing.

The ability to reprocess with new sets of terms comes from the OR'ing function; the result of each pass is OR'd with that of the preceeding passes.

$$F'(h^*) = F(h^*) + C_{res}$$

The memory is correctly filled at the end of the final pass.

10.4.4.2 More variables

Where there are more than 22 variables, the hardware design is limited. The limitation is due to the memory requirements. The design has not been extended at this time.

The memory requirements may be gauged by noting that:

for binary space:

$$2^{22} = 4,194,304 \text{ bits of storage}$$

for trinary space:

$$3^{14} = 4,782,969 \text{ bits of storage}$$

10.5 THE COVERAGE ALGORITHM

The coverage algorithm is a recently developed application for the parallel Boolean processor. It is represented in the last step of the logical instruments solution to minimization. (The paper defining the application in detail has not yet been published.)

The solution makes use of two concepts: (1) multilicity and (2) coverage. A special counter is also required for the processor that is capable of counting a 0, 1, or more-than-1 condition. Figure 10.17 presents the

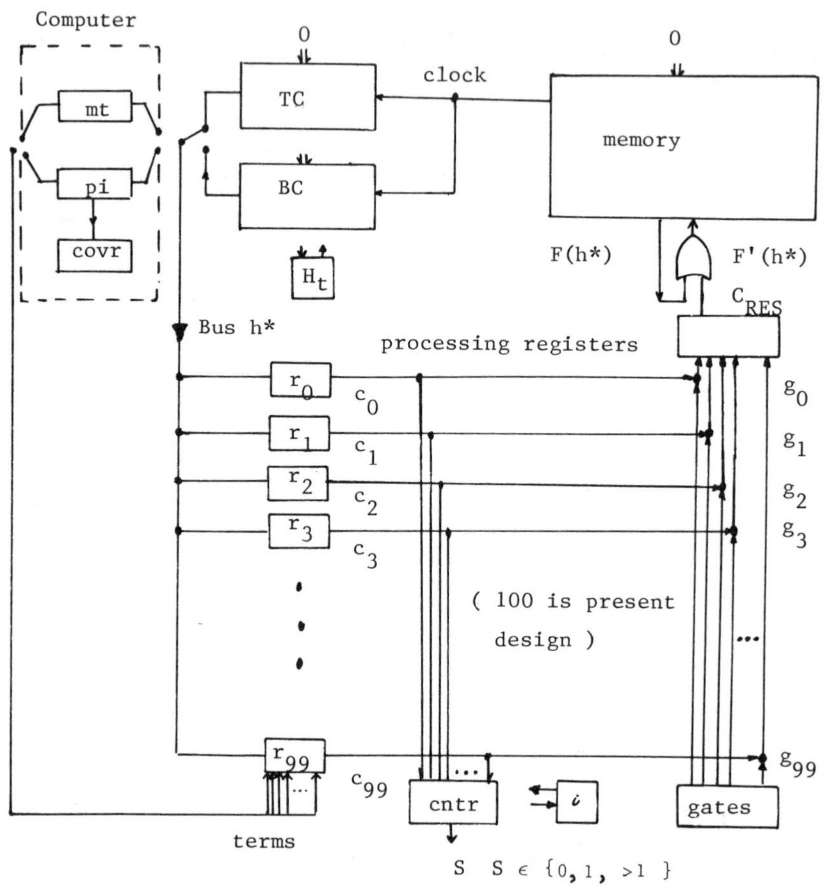

mt = minterms

pi = prime implicants

covr = coverage

TC = triadic counter $h_i^* \in \{ \bar{0}, 1, 2 \}$

BC = binary counter $h_j^* \in \{ 1, 2 \}$

Figure 10.17.   Block diagram - Parallel Boolean Processor.

modified block diagram of the processor.

Multiplicity is diagrammed in Figure 10.18. It is the production of a count of the number of prime implicants which cover a given minterm. The processing registers are loaded with the prime implicants found for the function. The clock bus represents the minterms. The result bits are formed by checking for NON3 as in the implicant, prime implicant, and existence function applications. A NON3 indicates coverage of the minterm on the bus by the prime implicant in the register. An essential prime implicant is found when the count of prime implicants forming NON3 with any minterm is 1. The sum condition must be recorded for each minterm.

Coverage is demonstrated in Figure 10.19. It is an algorithm to find the optimum $\Sigma\Pi$-form of the solution for a function. First, the processing registers are loaded with the minterms of the function and the prime implicants are sent down the bus. The prime implicants are sorted based on their status as essential or dominant prime implicants. A NON3 indicates that the prime implicant on the bus covers the minterm in the register. The prime implicant is recorded as belonging to the solution and all minterms forming non-3 with it are removed from the register.

The next step is to send the next prime implicant down the bus, repeating the above. This process is repeated until there are no more terms in the register.

This application is based upon the modified form of Theorem 3.3:

$$(h_j + h_j^* \neq 3 \text{ for every } j)$$
$$\rightarrow [(t_{h^*} = m_a) \rightarrow (m_a = t_h)]$$

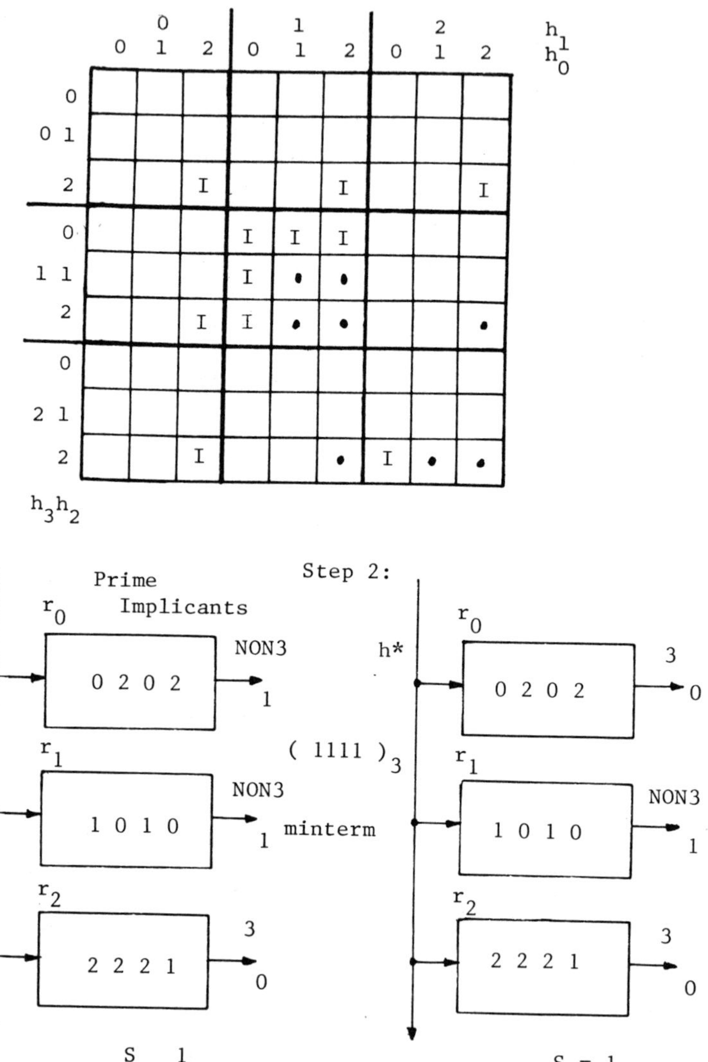

t_{0202}, t_{1010} covers t_{1212}
t_{2221} does not cover t_{1212}

t_{1010} essential prime implicant

Figure 10.18. Map and clock step detail of MULTIPLICITY.

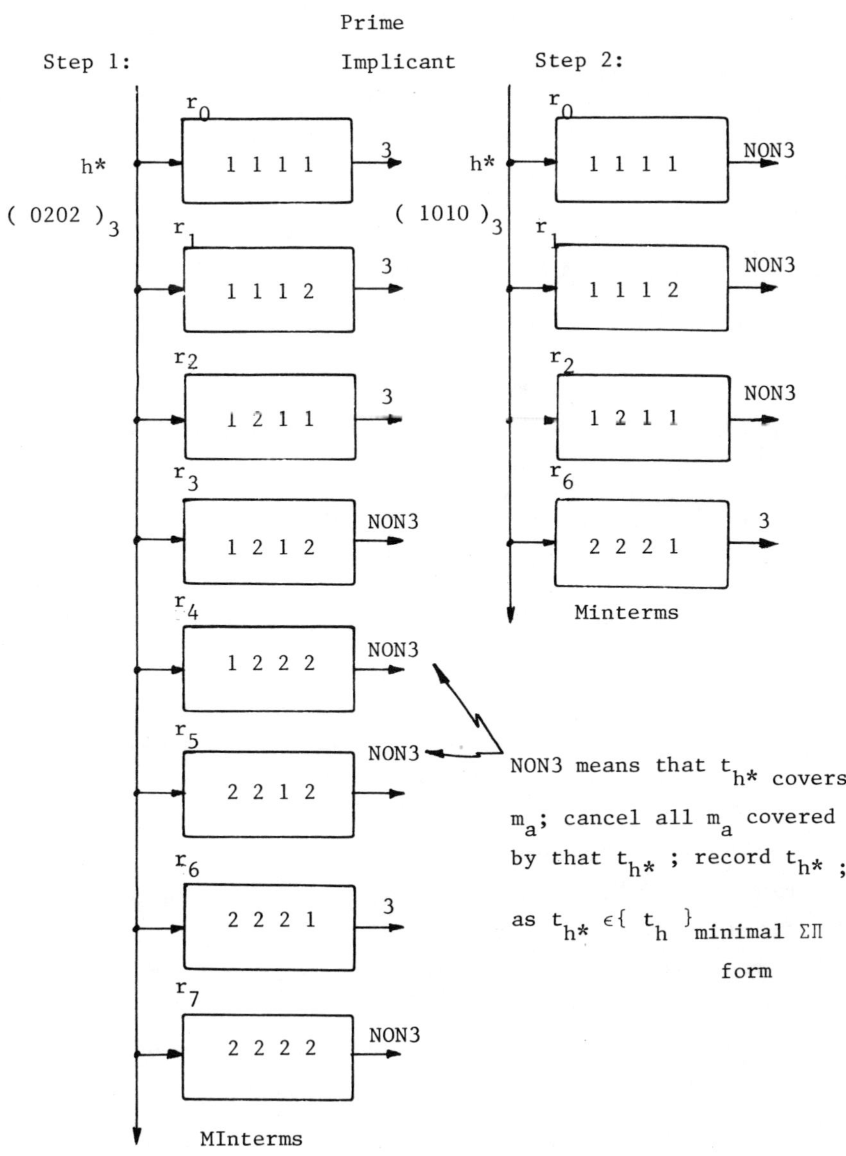

Figure 10.19. Clock step detail of COVERAGE.

10.6   THE BOOLEAN DIFFERENCE

The Boolean difference is defined as:

$$D_i F(x) = F(x_0, x_1, \ldots, x_i, \ldots, x_{n-1})$$
$$\veebar F(x_0, x_1, \ldots, \overline{x_i}, \ldots, x_{n-1})$$

or,

$$D_i F(x) = F(x_0, x_1, \ldots, 1, \ldots, x_{n-1})$$
$$\veebar F(x_0, x_1, \ldots, 0, \ldots, x_{n-1})$$

When $D_i F(x) = 0$, $y = F(x)$ is unconditionally independent of x_i.

When $D_i F(x) = 1$, $y = F(x)$ is unconditionally dependent on x_i.

When $D_i F(x) = G(x)$, $y = F(x)$ is conditionally dependent on x_i.

The literature previously published on the Boolean difference also defines operational properties for the difference.

The function is well documented in the literature and is used in some methods of minimal or optimum test set generation.

The Boolean difference has been derived for an example function in Figure 10.20 both by computation and by using a graphical map techique. The map technique shown uses a Marquand map; previous papers have used Karnaugh maps.

An alternative map approach, again using Marquand maps and the example is performed as follows:

1.   Map the function $y = F(x)$ (Figure 10.21a).

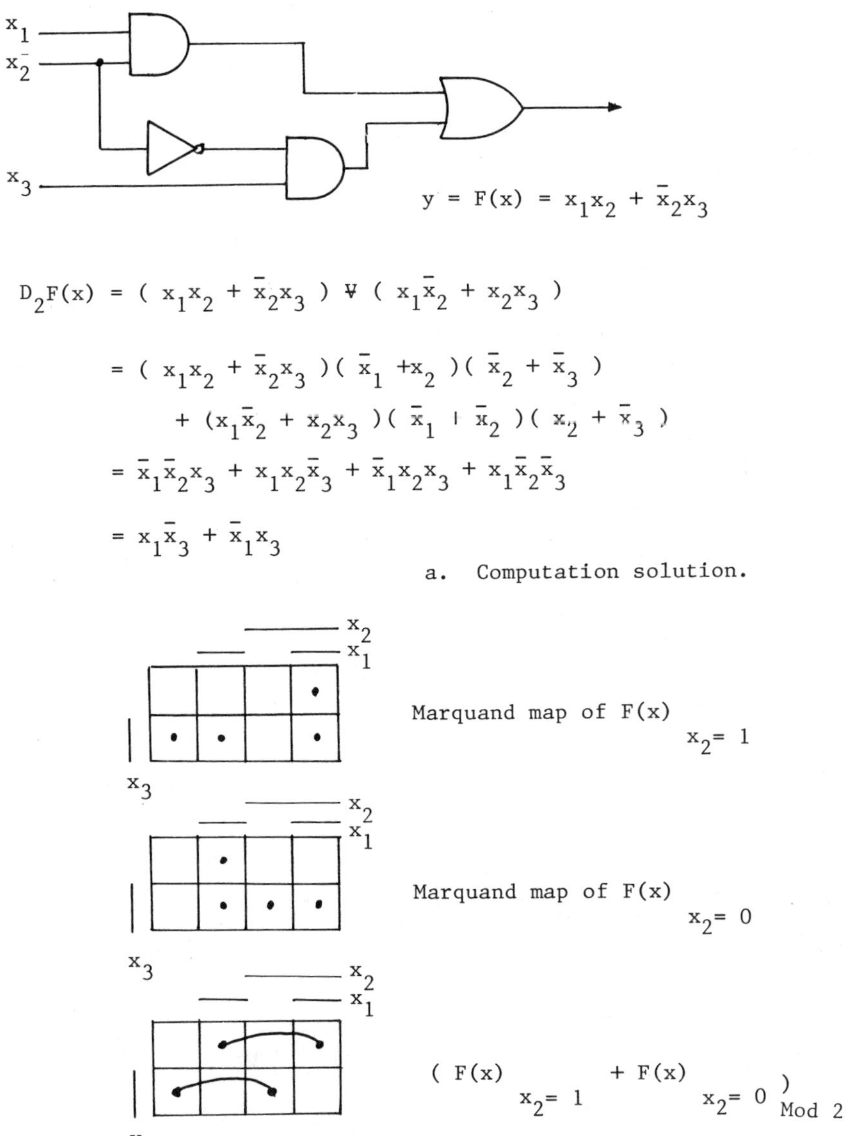

Figure 10.20. Boolean Difference by computation and mapping.

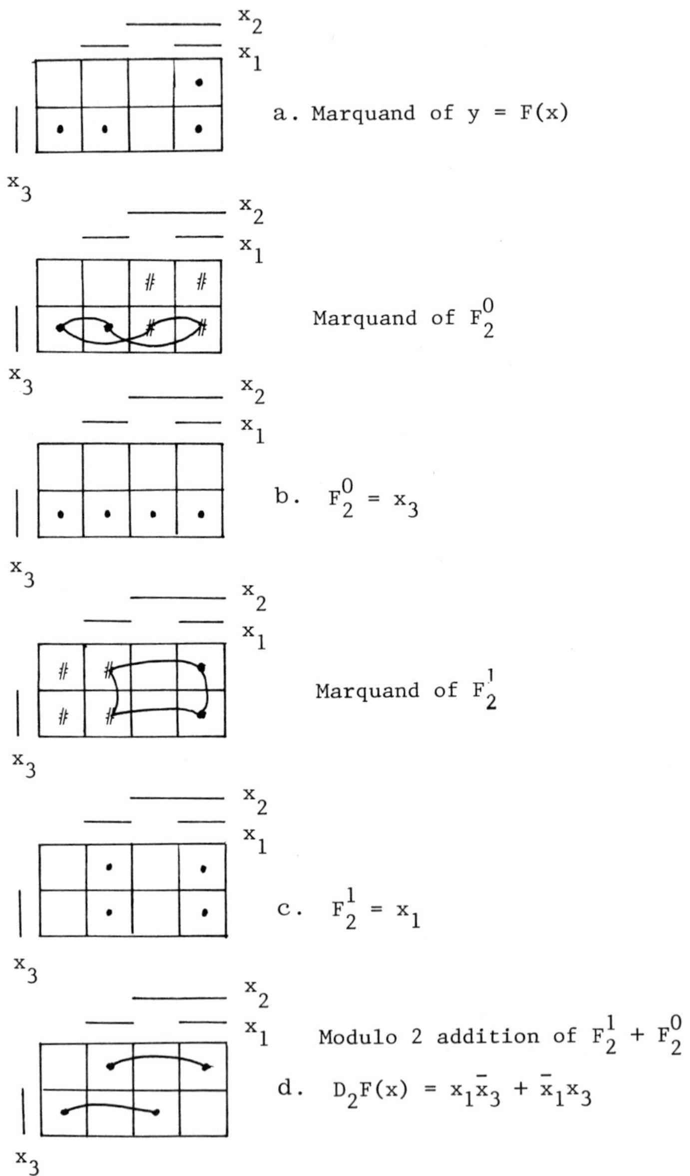

Figure 10.21. An alternative map approach.

2. To find $D_2F(x)$ computationally, we may define

$$D_2F(x) = F_2^1 \not\vee F_2^0$$

To find F_2^0 consider all points where $x_2 = 1$ to be "don't care"s and minimize the resulting function for F_2^0 (Figure 10.21b).

3. To find F_2^1 consider all points where $\overline{x_2} = 1$ to be "don't care"s and minimize for F_2^1 (Figure 10.21c).

4. The two functions, F_2^0 and F_2^1, must be exclusive OR'd together to find $D_2F(x)$, which is accomplished by the modulo 2 addition of their maps (Figure 10.21d).

This approach was suggested by Svoboda. The choice of one of these methods over the other is arbitrary, dependent upon the ease of switching x_i with $\overline{x_i}$ and remapping, versus minimization with "don't care" masks. The choice of which will be automated has not been made. The map approach was originally intended for manual use for cases where n was reasonably small.

A third approach involves the use of "links" (defined in "Fault Detection through Parallel Processing in Boolean Algebra", Ph.D. Thesis, UCLA, 1976, by D. E. White). A "link" exists if, for all x_j except the x_i in question held constant, a change in x_i produces a change in y. (Figure 10.22a.) This is best seen using the Existence Function map (Figure 10.22b).

The procedure:

1. Using the Existence Function map, identify all links for the variable x_i.

Find all links for the variable x_i
for which the difference is being
sought. This is clearer on the
Existence function map.

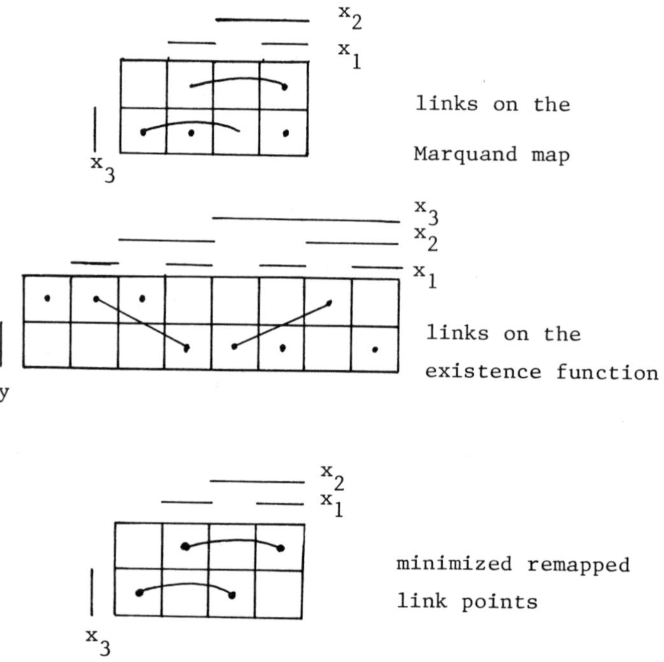

The minimized function found from
the points on the Marquand map is
the Boolean Difference being sought.

Figure 10.22. Solution with links and the existence function map.

    2.   On a new Marquand map, map only those points bound by links for the variable x_i (Figure 10.22c).

    3.   Minimize the function represented by these points to find $D_i F(x)$.

Given these methods, it is desired to define algorithms for the parallel Boolean processor. The present design limitation will limit these algorithms to cases where $n \leq 22$.

Algorithms/routines which already exist for the parallel Boolean processor are referenced but are not detailed here.

The algorithm for the Marquand map approach first described:

    1.   Load a 2^n Boolean space with $F(x)\big|_{x_i=1}$.

    2.   Load a second 2^n Boolean space with $F(x)\big|_{x_i=0}$ (software instruction).

    3.   Using software to add these maps modulo 2, load a 3^n triadic space with the result map.

    4.   Run the implicant algorithm.

    5.   Run the prime implicant algorithm. The result is $D_i F(x)$.

The storage required for the maps (spaces) is $2^{n+1} + 3^n$ bits. The algorithm is illustrated in Figure 10.23.

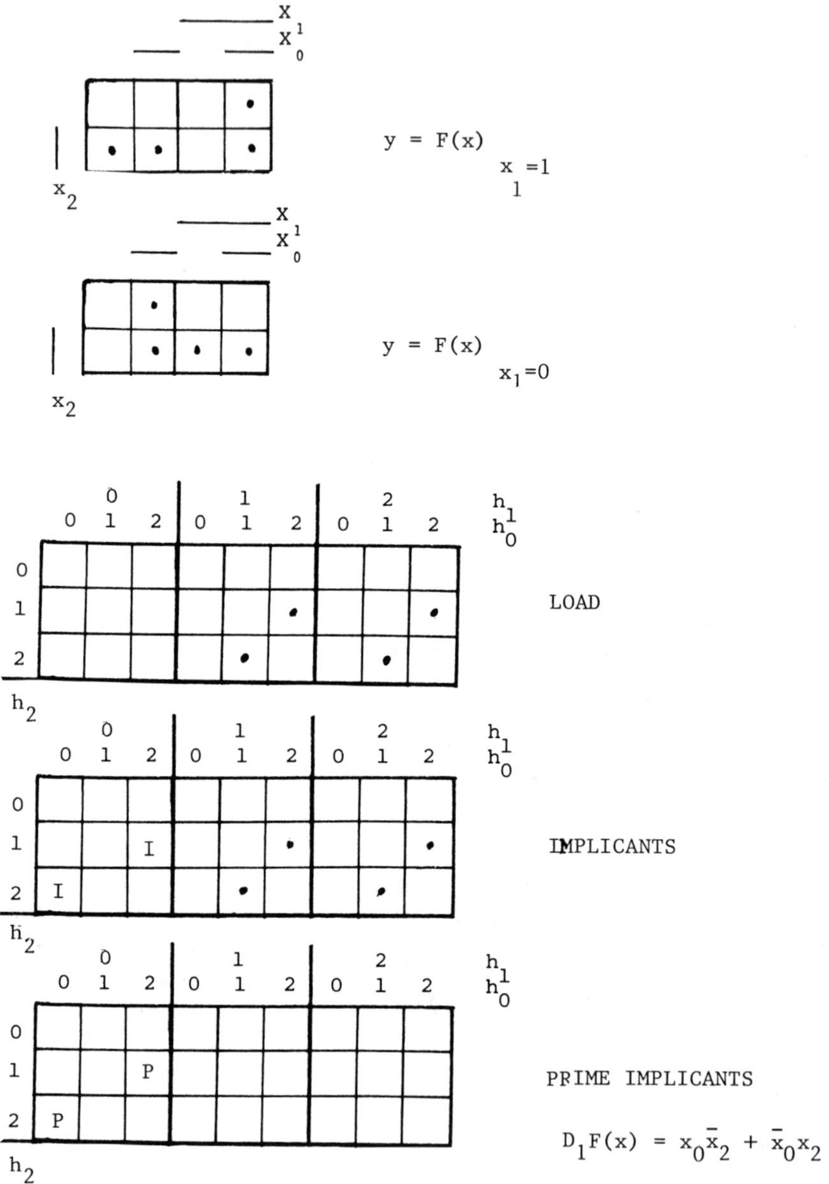

$$D_1 F(x) = x_0 \bar{x}_2 + \bar{x}_0 x_2$$

Figure 10.23. Parallel Boolean Processor algorithm for implementation of map approach.

INTRODUCTION TO THE PARALLEL BOOLEAN PROCESSOR

The algorithm for the link approach is as follows:

1. Load the existence function $E(y)$ into a 2^{n+1} Boolean space.

2. Find the links for the variable x_i (this algorithm is needed for the test sequence generation and is therefore considered to exist).

3. Load a 3^n triadic space with the linked points of $E(y)$.

4. Run the implicant algorithm.

5. Run the prime implicant algorithm. The result is $D_i F(x)$.

The storage required for the maps (spaces) is $2^{n+1} + 3^n$ bits.

A simple example has been used throughout this description. The algorithms have, however, been succesfully applied to a number of examples of varying complexity, including several published in the literature. A more complex example is illustrated in Figures 10.24 (Marquand map method) and 10.25 (link method).

These algorithms have been extended for the generation of the Boolean difference of y with respect to two variables:

(both)

$$D_{ij}F(x) = F(x_1, \ldots, x_i, \ldots, x_j, \ldots, x_n)$$
$$F(x_1, \ldots, \overline{x_i}, \ldots, \overline{x_j}, \ldots, x_n)$$

(either)

$$D_{i \vee j}F(x) = D_i F(x) + D_j F(x)$$

(either or both)

$$D_{i+j}F(x) = D_{ij}F(x) + D_{i \vee j}F(x)$$

$$y = F(x) = x_1 x_2 x_3 + x_1 x_2 x_4 + \bar{x}_1 x_2 x_5$$

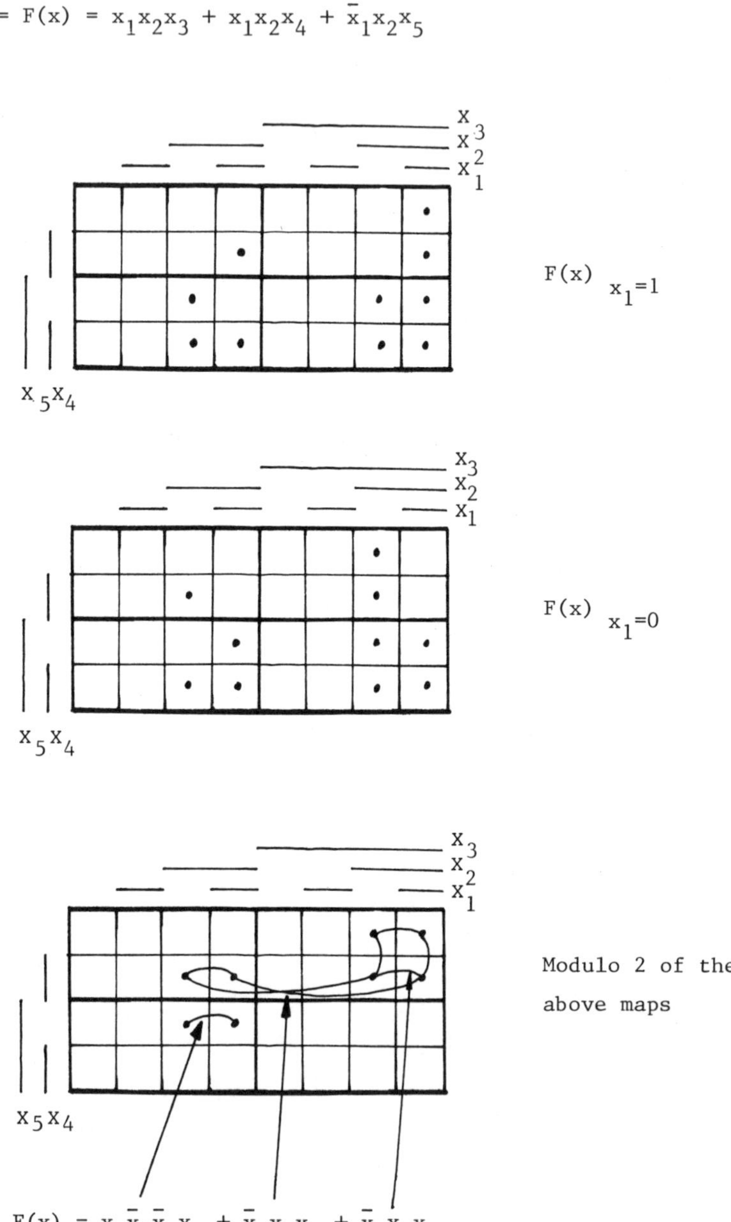

$$D_1 F(x) = x_5 \bar{x}_4 \bar{x}_3 x_2 + \bar{x}_5 x_4 x_2 + \bar{x}_5 x_3 x_2$$

Figure 10.24. Parallel Boolean Processor algorithms example.

By computation:

$$D_1 F(x) = D_1 [x_1 x_2 x_3 + x_1 x_2 x_4 + \bar{x}_1 x_2 x_5]$$

$$= (x_1 x_2 x_3 + x_1 x_2 x_4 + \bar{x}_1 x_2 x_5) \; \forall \; (\bar{x}_1 x_2 x_3 + \bar{x}_1 x_2 x_4 + x_1 x_2 x_5)$$

$$= (x_1 x_2 x_3 + x_1 x_2 x_4 + \bar{x}_1 x_2 x_5) (x_1 + \bar{x}_2 + \bar{x}_3) (x_1 + \bar{x}_2 + \bar{x}_4) (x_1 + \bar{x}_2 + \bar{x}_5) + (\bar{x}_1 x_2 x_3 + \bar{x}_1 x_2 x_4 + x_1 x_2 x_5) (\bar{x}_1 + \bar{x}_2 + \bar{x}_3) (\bar{x}_1 + \bar{x}_2 + \bar{x}_4) (x_1 + \bar{x}_2 + \bar{x}_5)$$

$$= x_1 x_2 x_3 \bar{x}_5 + x_1 x_2 x_4 \bar{x}_5 + x_1 x_2 \bar{x}_3 x_4 x_5 + x_1 x_2 x_3 \bar{x}_4 \bar{x}_5$$

$$+ \bar{x}_1 x_2 \bar{x}_3 \bar{x}_4 x_5 + x_1 x_2 \bar{x}_3 \bar{x}_4 x_5 + \bar{x}_1 x_2 x_3 \bar{x}_5 + \bar{x}_1 x_2 x_4 \bar{x}_5$$

$$+ \bar{x}_1 x_2 \bar{x}_3 x_4 \bar{x}_5 + \bar{x}_1 x_2 x_3 \bar{x}_4 \bar{x}_5$$

$$= x_2 \bar{x}_5 (x_3 + x_4) + x_2 x_5 \bar{x}_3 \bar{x}_4$$

Ref: Sellers, et. al. "Analyzing Errors with the Boolean
Difference." IEEE Trans., C-17(4):676–683, July 1968;
correction C-20(11);1245–1251, November 1971.

Figure 10.24. (con't)

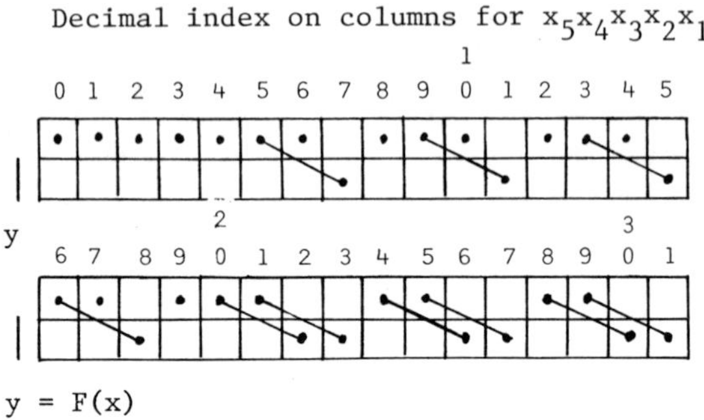

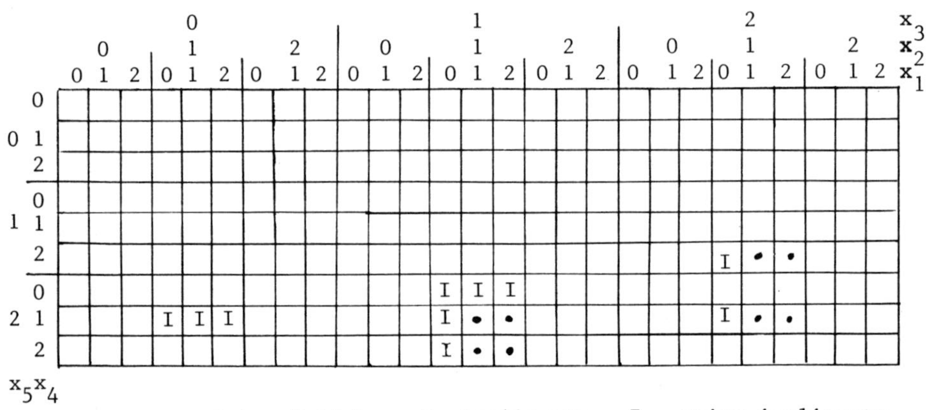

$$D_1F(x) = x_5\bar{x}_4\bar{x}_3x_2 + \bar{x}_5x_4x_2 + \bar{x}_5x_3x_2$$

Figure 10.25. The solution with links.

and may be extended to all n variables. No work has yet been done to extend them to the partial Boolean difference.

A final example is given in Figure 10.26.

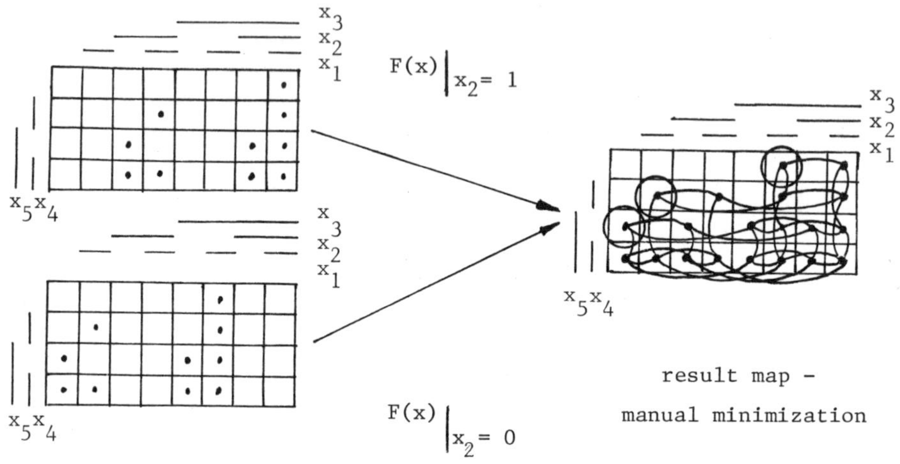

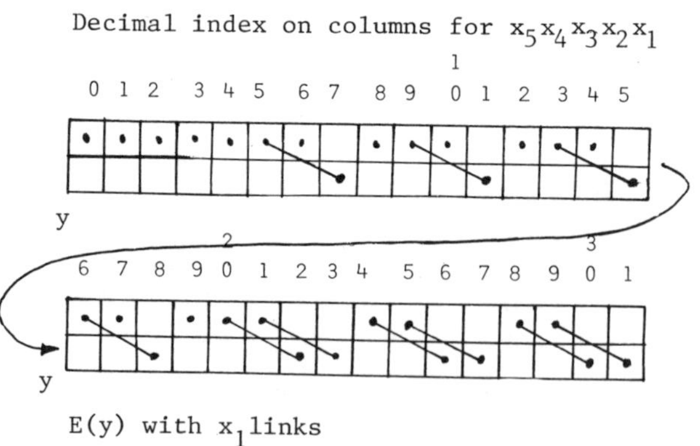

Figure 10.26.   Another example difference from Sellers.

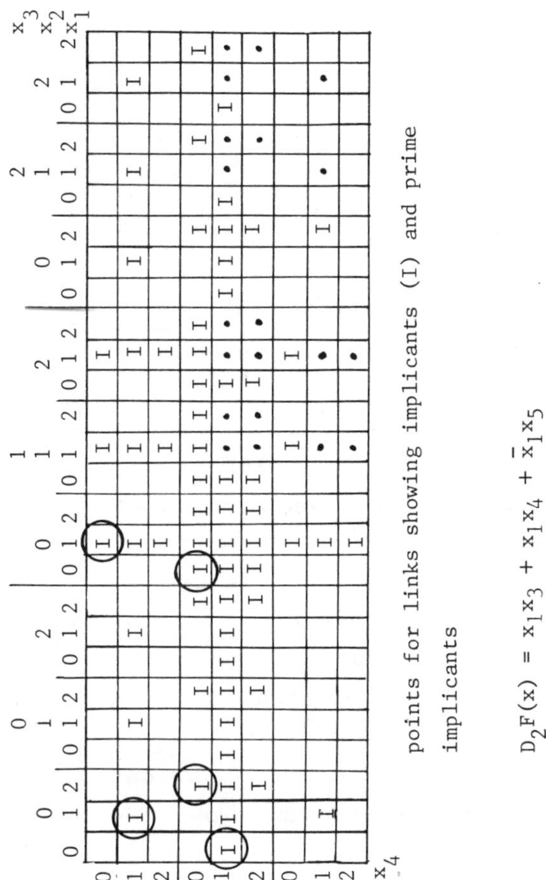

Figure 10.26. (con't)

Chapter 11
Designing with MSI-LSI

11.1 INTRODUCTION

Once a set of design constraints have been estab-
lished, there are several options available to the
designer for the implementation. Circuits built with
SSI-level logic may be faster, may be minimal in parts
count, may be irreducible and, therefore, may be more
testable than those built using MSI or LSI devices. SSI
level designs take a relatively longer time to develop,
debug and document. It is acknowledged to be the most
difficult . The criteria that determine the desirability
of doing an implementation with SSI-level logic are: 1) the
required speed is faster than would be available with
any other implementation; and 2) the anticipated production
volume justifies the expense of doing an SSI-level design.

Where total testability is not a major concern, as
is true for most commercial-level applications, MSI
design with multiplexers and function blocks is attractive. While multiplexers are non-minimal implementations
of their output functions, which constrain their testability, they are easier to design with and allow reduction
in board space. Speed and power requirements are the
determining factors in choosing to use multiplexers and
other MSI function blocks.

DESIGNING WITH MSI-LSI

Programmable multiplexers (PMUX's), gate arrays
(PGA, CGA), logic arrays (PLA, FPLA) and array logic
devices (PAL) are available now in a number of configurations and sizes. Most of these are intended to assist
in reducing the parts count for implementations of combinational functions. Some of the PALs have registered
outputs and feedback paths and are for use in sequential
logic implementation. All of these devices are contri-
buting toward modular hardware designs.

Sequential control functions may be implemented with
microprograms using PROM/ROM's with the more complex controls using a microprogram sequencer. Higher-speed control functions are being implemented using bipolar bit-
slice devices (such as the AMD AM2910) while less speed-
restricted applications use one or more of the micropro-
cessor/microcomputers (such as the Intel 8085). All of
these devices are considered to be LSI and all of them
require a software investment. The bit-slice devices
require microprogramming, which may be accomplished with
a psuedo-assembly-level language and a development system.
The fixed-instruction-set microcomputers are programmable
in assembly-level or higher-level languages, with some of
the new devices to be programmable in PASCAL or a similar
algorithmic language.

As the implementation shifts from SSI to LSI, software design techniques, specifically modularity and structured programming, begin to become mandatory. Testing is
also no longer a hardware checkout operation, but requires
software diagnostic packages. One advantage of LSI is the
feasibility of putting the system diagnostic routines on-
board the PROM's, to simplify design debug and field testing.

Software development costs in computer systems have
far exceeded those of the hardware development. This

trend is being repeated in LSI designs with the microprogramming development costs overriding the hardware
costs.

11.2 SSI DESIGN

To develop a minimal or optimal two- to three-level
NAND logic circuit, with the third level for inversion
of the input variables, the equation of the function y
must be solved for its minimal $\Sigma\Pi$-form. This is accomplished using the APL function

$$\text{MINIMA (Z 1)}$$

which produces the desired expression.

To develop a minimal or optimal two- to three-level
NOR logic circuit, the equation of the function y must
be solved for its minimal $\Pi\Sigma$-form. This is accomplished
using the APL function

$$\text{MINIMA (}\underline{Z}\text{ 1)}$$

which produces the desired expression.

Conversion techniques exist for NOR to NAND and from
NAND to NOR without restarting from the original expression of the function y (refer to any text for a beginning
course in logic design); however, the resulting network
is not guaranteed to be minimal. Where an expression is
factorable to produce terms of the form (a $\bar{b}$ + $\bar{a}$ b), EXOR
gates may be used to reduce gate count and to simplify
implementation.

Fan-in requirements can be achieved by factoring the
minimal $\Sigma\Pi$- or $\Pi\Sigma$-forms. The fan-in requirements present
in early SSI designs are more relaxed today with allowan-
ces of 8 or more inputs per gate. Fan-out requirements
are generally alleviated by using buffer-drivers. Present
designs use a conservative limit of 7 loads per output
for devices rated at 10 loads.

11.3 GATE VERSUS CONNECTION MINIMIZATION

Muruga and Lai (Muruga and Lai, "Minimization of Logic Networks under a Generalized Cost Function", _IEEE Trans._, Sept. 1976, pp. 893-907) reported on the calculations of the minimal networks of NOR gates for all functions or 3 or fewer variables and for some of the functions of 4 variables. For the 77 P-equivalence nontrivial class representative functions of 3 or fewer variables, there are only two functions for which the minimal networks under GCM (gate reduction emphasis) and CGM (connection reduction emphasis) differ. They are:

$$Y_1 - X_2 \oplus X_1 \oplus X_0$$

and

$$Y_2 = X_2 X_1 X_0 + \bar{X}_2 (\bar{X}_1 + \bar{X}_0)$$

For these two functions, the minimal $\Pi\Sigma$-form was investigated and found to give the same number of gates as does the GCM reduction; more connections than does the CGM version; and fewer stages (gate levels and therefore gate delays).

The GCM reduction reduces the gate count by one over the minimal $\Pi\Sigma$-form, but increases the connections by one. In one case there is an added stage delay. The minimal $\Pi\Sigma$-forms of the two functions are shown in Figure 11.1a and 11.1b. A table comparing the various implementations in terms of number of gates, number of connections and number of levels is shown in Figure 11.1c.

Among 312 representative functions of P-equivalence classes of 4 variables requiring at most 5 NOR gates under GCM, 26 functions were found to have different minimal networks under CGM design constraints.

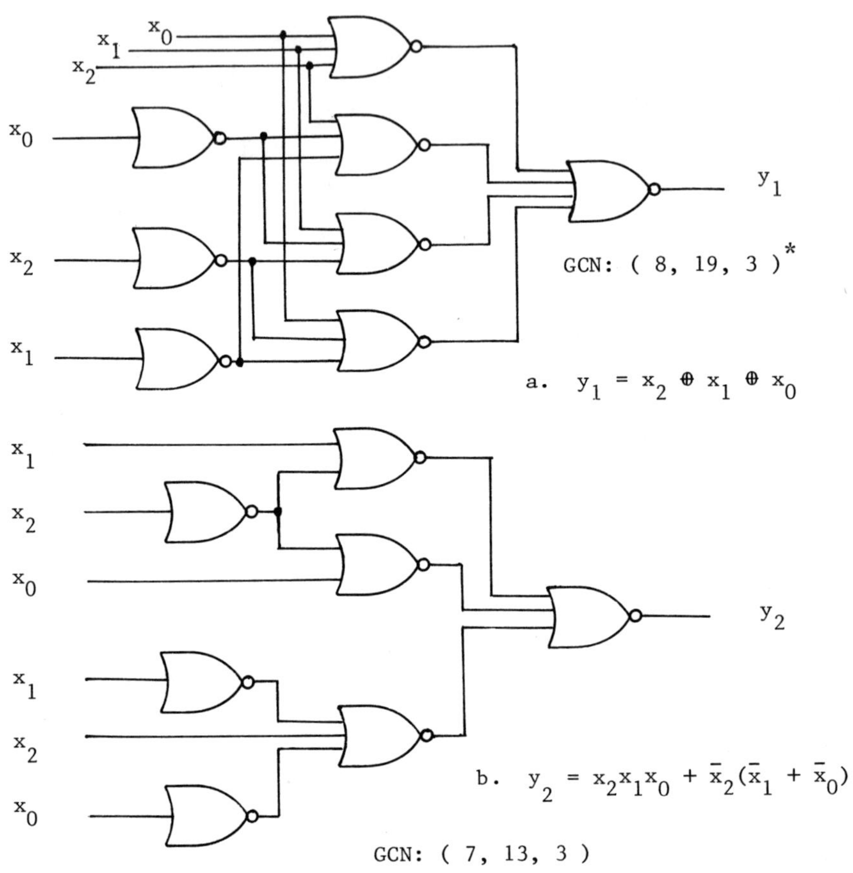

a. $y_1 = x_2 \oplus x_1 \oplus x_0$

b. $y_2 = x_2 x_1 x_0 + \bar{x}_2 (\bar{x}_1 + \bar{x}_0)$

Minimal ΠΣ-forms of y_1 and y_2

function	minimal	GCM	CGM
y_1	(8, 19, 3)	(7, 20, 4)	(8, 16, 6)
			(8, 16, 5)
			(8, 16, 5)
y_2	(7, 13, 3)	(6, 14, 3)	(7, 12, 4)
			(7, 12, 5)

Figure 11.1.   Minimal ΠΣ forms of y_1 and y_2.

* (gates, connections, levels)

Of the 26 functions:

(1) 4 of the functions had GCM and CGM minimal
     networks where the number of gate levels or
     stages were equal. In 22 cases, the number
     of levels was at least one higher for CGM
     design constraints.

(2) 7 of the functions implemented in CGM had a
     gate count higher than the GCM version by 1.
     In the rest the gate count was higher by 2.

(3) 8 of the functions implemented in CGM had 2
     fewer connections than the GCM versions of
     those functions. The rest had a connnection
     count that was lower by 1.

(4) For some of the functions, a gate count for
     the GCM version lower than the minimal $\Pi\Sigma$-
     form version was obtained by the introduction
     of redundancy which in turn reduced testabi-
     lity.

(5) For some of the functions, a gate count re-
     duction was accomplished through the use of
     a factored, equivalent form which added one
     gate level.

Their conclusion was to use the gate criteria in minima-
zation and design (they were concerned specifically with
chip area in their paper). Our conclusion is to use the
minimal $\Pi\Sigma$- and $\Sigma\Pi$-forms, factoring where possible for
reduction but maintaining irredundancy.

The length of time available to design the network
is an overriding constraint in all cases. For the mini-
mal $\Sigma\Pi$- and $\Pi\Sigma$-forms, the APL function MINIMA is available as a design tool. Figure 11.2 examines three implementations of the function Muruga and Lai indexed as
EBFF. (Their index uses HEX notation to fill the rows
of a 4-variable Marquand map.) It presents the minimal

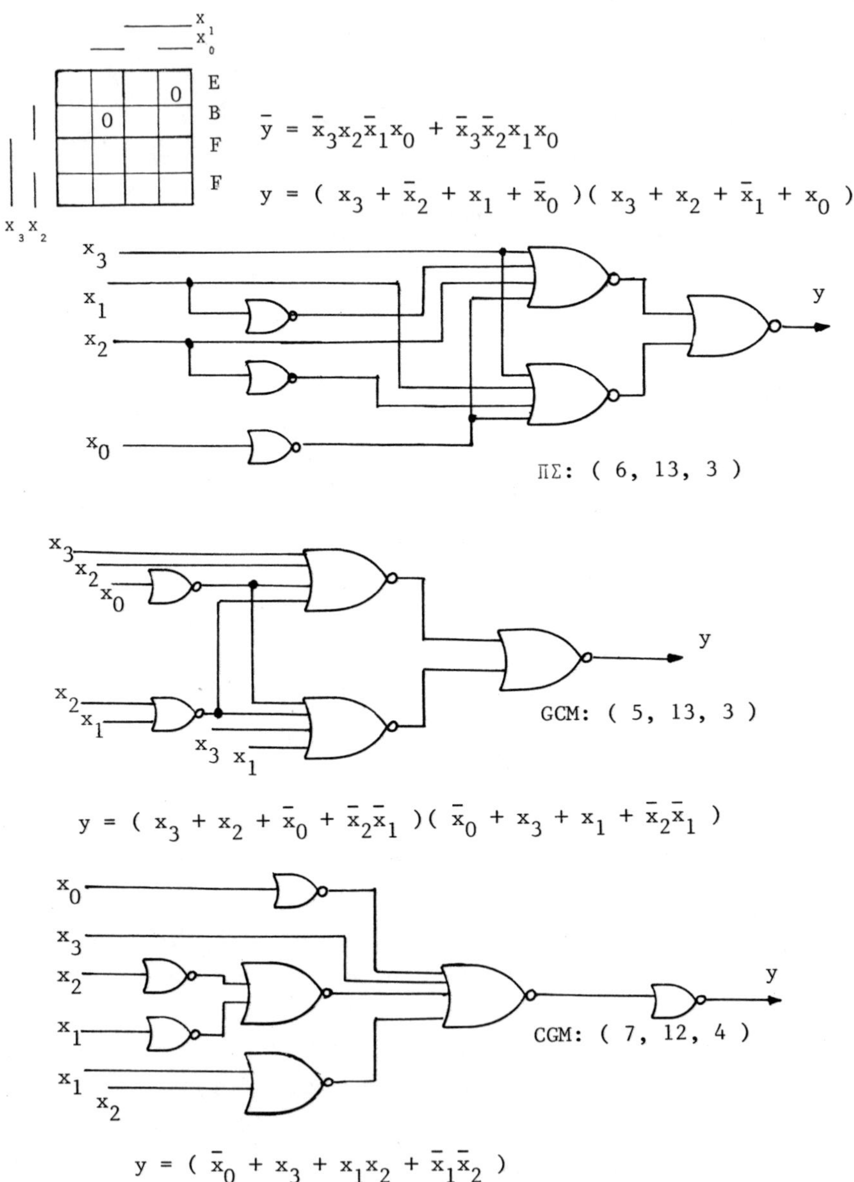

$$\bar{y} = \bar{x}_3 x_2 \bar{x}_1 x_0 + \bar{x}_3 \bar{x}_2 x_1 x_0$$

$$y = (x_3 + \bar{x}_2 + x_1 + \bar{x}_0)(x_3 + x_2 + \bar{x}_1 + x_0)$$

$$\Pi\Sigma: (6, 13, 3)$$

$$GCM: (5, 13, 3)$$

$$y = (x_3 + x_2 + \bar{x}_0 + \bar{x}_2\bar{x}_1)(\bar{x}_0 + x_3 + x_1 + \bar{x}_2\bar{x}_1)$$

$$CGM: (7, 12, 4)$$

$$y = (\bar{x}_0 + x_3 + x_1 x_2 + \bar{x}_1\bar{x}_2)$$

Figure 11.2. Comparison of SSI implementations.

$\Pi\Sigma$-form, a GCM (minimal gate) version, and a CGM (minimal connection) version of the function.

For the two functions of three variables Y_1 and Y_2 described earlier, the SSI implementation of Y_1 is improved by using EXOR gates. A comparison between the implementation of the minimal $\Pi\Sigma$-form of y in NOR gates and the EXOR implementation is shown in Figure 11.3.

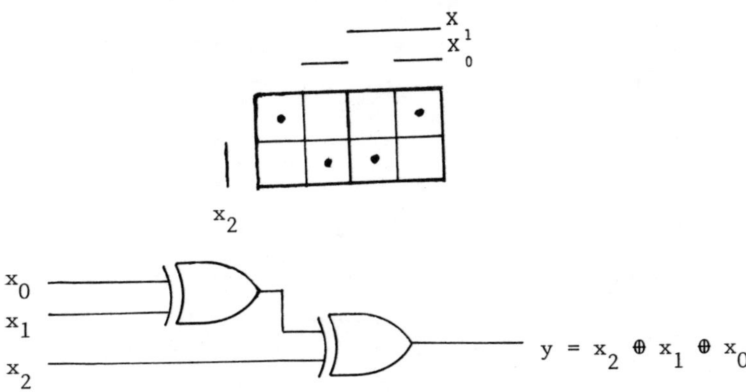

Item	Minimal $\Pi\Sigma$ NOR	EXOR
Cost	0.80	0.40
Levels	3	2
Packages	4	1
Time (speed)	32ns	28ns
Power	56mW	150mW

Figure 11.3. An EXOR implementation of y_1 and comparison to the minimal $\Pi\Sigma$ form. (Data from available data sheets)

11.4   MSI DESIGN

The decision to use SSI or MSI is dependent upon the complexity of the function being implemented and upon design constraints such as timing, board space, and power consumption.

Multiplexer chips for 3, 4, and 5 variables are easy to use and input requirements can be read directly from the Marquand map of the function. (Karnaugh maps require permutation of certain columns to achieve a column-to-input order and are therefore not recommended.) They are available in some cases with true and complemented outputs and have 1, 2 or 3 chip select or enable pins, providing design flexibility.

Multiplexers are formed internally from two-level logic which makes them competitive in speed with SSI gate implementations. In some cases they may even be faster than SSI due to the reduction of interpackage time delays.

The use of multiplexers reduces package count at the cost of increased power consumption per package. The other factors that must be considered are: (1) The number of connections; (2) the loading on variables; (3) the loading on power and ground. The use of multiplexers may or may not render the design "flexible", i.e., able to be altered with changes in requirements and specifications. Any function implemented in multiplexers is not necessarily minimal. Coverage will be column-minimal at the cost of possible increased loading. Figure 11.4 presents the one-of-four multiplexer in detail and an example derivation of input functions using a Marquand map.

Figure 11.5 presents the multiplexer implementations of Y_1 and Y_2 and a table summarizing the differences between the multiplexer, NOR gate and EXOR implementations.

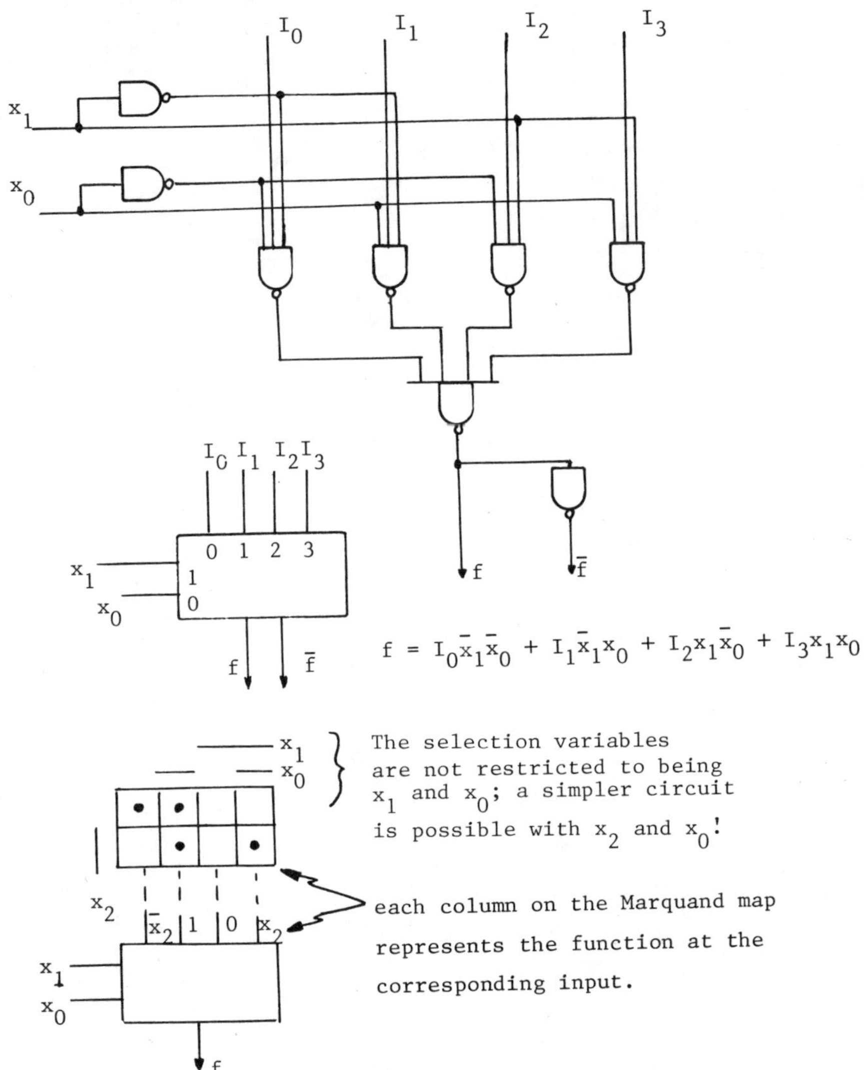

Figure 11.4.   One of four multiplexer.

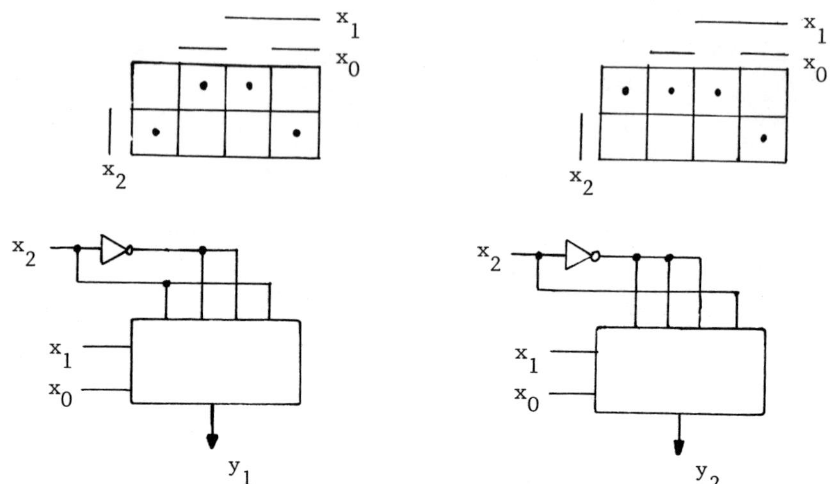

Comparing the implementations:

Time delay:	21 ns	faster than NOR gates
Power:	200 mW	more power, more heat
Package vount:	1	same as for EXOR version
Cost:	$1.50	(approx) more expensive

(compare to Figure 11.3.)

Figure 11.5.   Implementation of y_1 and y_2 with multiplexers.

Multiplexer sizes are 1-of-2, 1-of-4, 1-of-8 and 1-of-16 (an oversized chip). When a function larger than five variables is to be implemented, multiplexers may be cascaded. The 4-variable function EBFF is shown in Figure 11.6a, implemented using an EXOR gate to improve the connection count (8 instead of 12 for the same number of gates) at the expense of an added level and its delay. Figure 11.6b shows the clean simplicity of a multiplexer implementation of the same function and the associated Marquand map.

Multiple-output problems are designed faster using multiplexers. The problem discussed in Chapter 8 and shown in Figure 8 is shown implemented with multiplexers in Figure 11.7. Included is a table comparing: 1) the individual output function SSI version; 2) a reduced multiple-output SSI version, and 3) a multiplexer version. The multiplexer version in this case runs slightly faster, has fewer connections, uses less board space, can be designed relatively faster, and is easily debugged, all at a cost of 2 to 4 times the power consumption. The choice is dependent upon the design constraint of specified allowable power consumption and heat dissipation.

As more variables are introduced, choices in which multiplexers to use are added. The six-variable example discussed in Chapter 8 is implemented using a 1-of-16 MUX and again using 1-of-8 MUX's in Figure 11.8. As the columns become "deeper" (number of variables in the input columns), minimization techniques are applied on a column-by-column basis (in reality, the map is listed as a set of 16 small function maps), as shown in Figure 11.8a.

The map of the function, its equation, and a table comparing the three implementations -- SSI and the two MUX versions -- are presented in Figure 11.9.

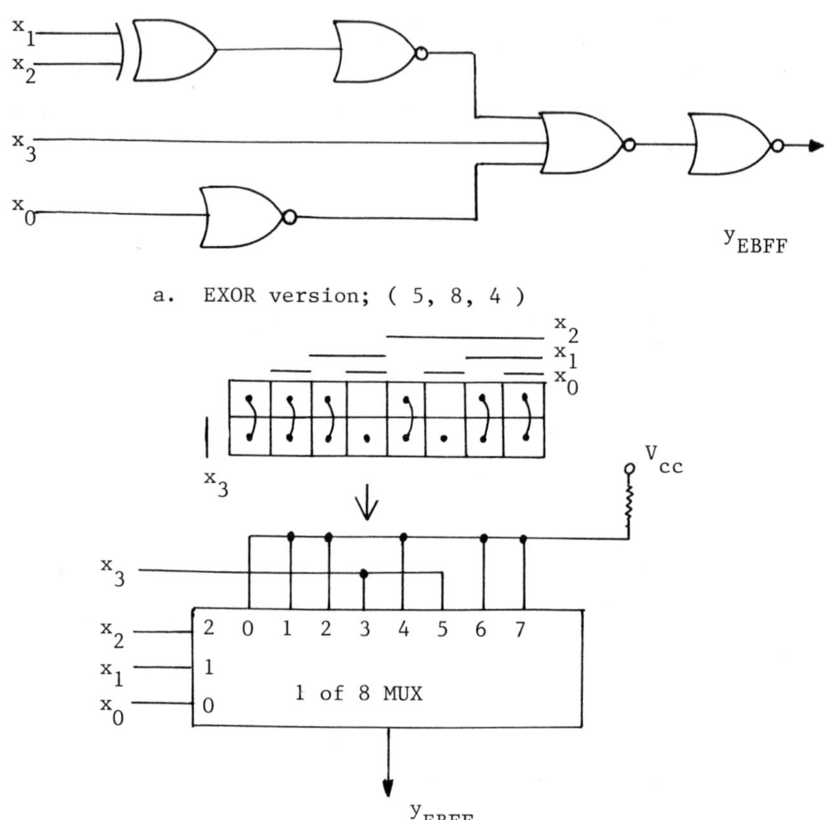

a.   EXOR version; (5, 8, 4)

b.   Multiplexer version; (1, 11, 1)

Figure 11.6. The four variable function y_{EBFF} from Muruga and Lai's paper.

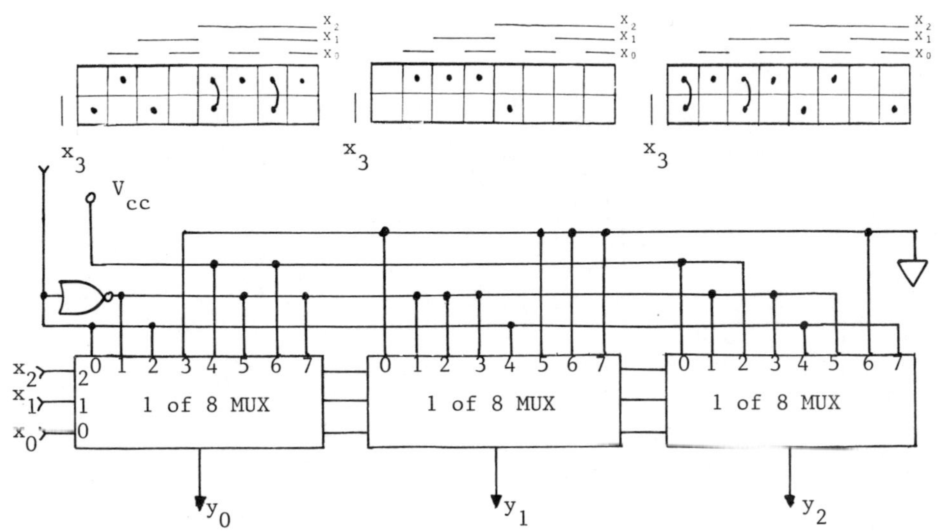

Multiple output problem implementation comparison

version	gates	packages	connec-tions	spares
SSI	17	6	43	0
Mult. Out.	15	6	38	3 gates
MUX	3 MUX +1	4	34	5 inverters

(version)	levels	power mW	speed ns
SSI	3	170	30
Mult. Out.	3	150	30
MUX	2	685	18
		445	26

Figure 11.7. Multiplexer implementation of the multiple output problem.

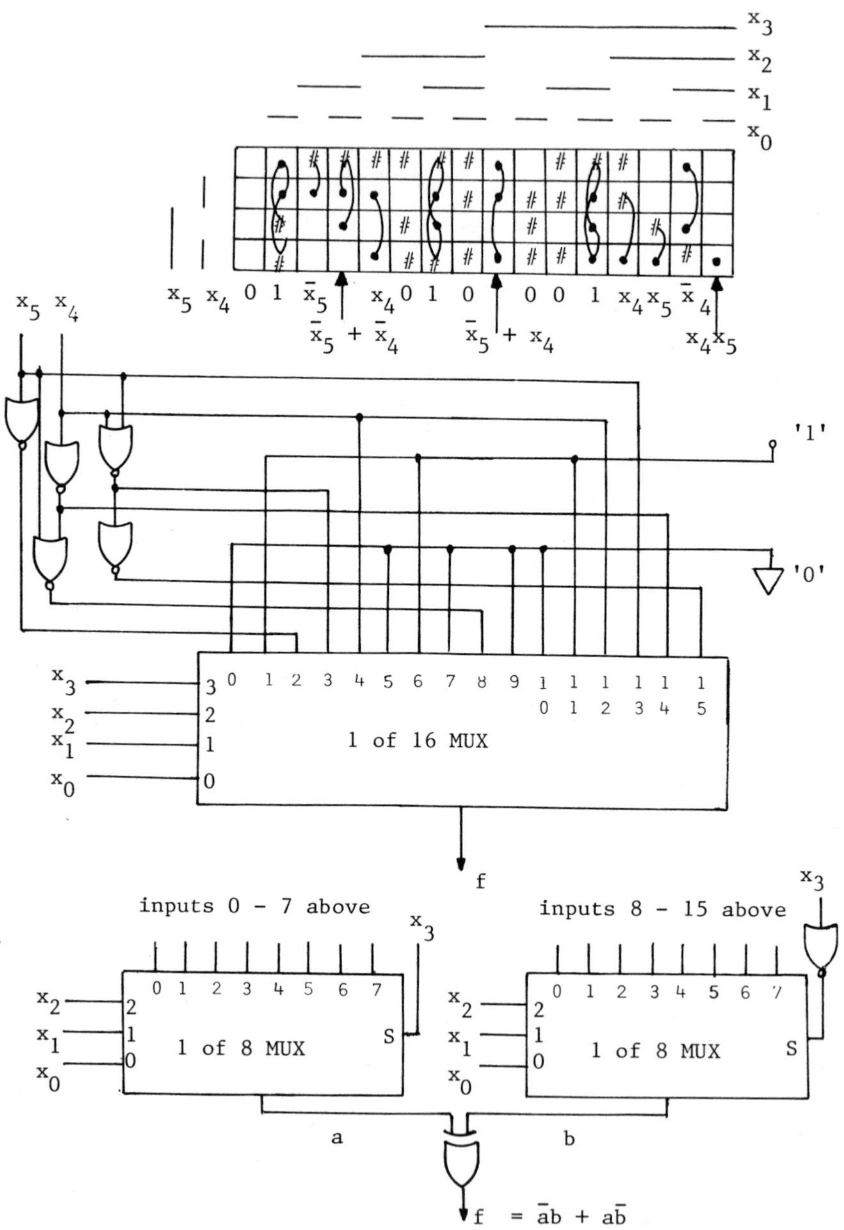

Figure 11.8.   Svoboda's six-variable example done with multiplexers.

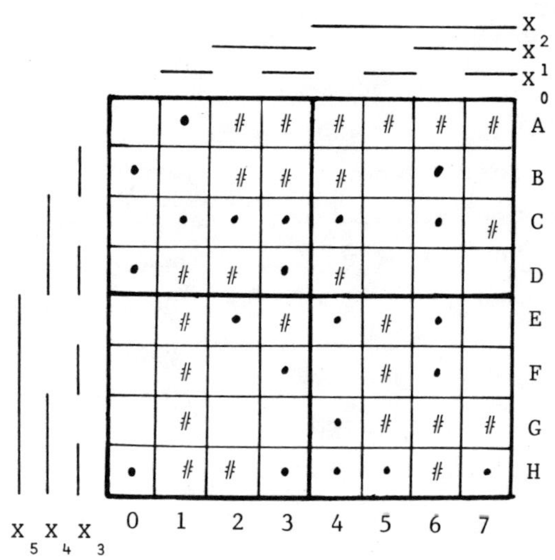

$$y = \bar{x}_5 x_3 \bar{x}_1 x_0 + \bar{x}_4 \bar{x}_3 x_1 \bar{x}_0 + \bar{x}_4 x_2 x_1 \bar{x}_0 + \bar{x}_5 \bar{x}_3 \bar{x}_2 x_0 + \bar{x}_5 \bar{x}_2 x_1$$

$$+ \bar{x}_3 x_2 \bar{x}_0 + x_5 x_3 \bar{x}_2 x_0 + x_5 x_4 x_3$$

Version:	stages	connec- tions	pack- ages	time	power	approx. cost
SSI (NAND)	3	43	6	30 ns	220 mW	1.20
MSI, 1 of 8 MUX, NOR, EXOR	4	37	4	51 ns	500 mW	2.60
MSI, 1 of 16 MUX, NOR, INV	4	28	3	41 ns	316 mW	1.80

(data from data sheets, numbers are typical values)

Figure 11.9. Comparison of implementations of the six-variable example.

217

11.5 LSI DESIGN TECHNIQUES

There are in existence a number of microprogrammable devices which may be used to reduce parts count and, therefore, required board space. A few examples of the more common of these are called out in Table 11.1. Programmable devices, with the exception of ROM's, are effective in design situations where the number of input variables is large and the number of active logic states is small -- that is, in cases where the logical map of the function is sparsely populated.

DEVICE	EXAMPLE PRODUCT
Programmable multiplexers	Raytheon 29693 PMUX (10 inputs, 1 term/MUX input, 4 1-of-8 MUXs)
FPLA	Signetics N82S101/10 (16 inputs, 48 terms, 8 outputs, 50 ns.)
PLA	National DM 75 7516 (19 inputs, 70 terms, 8 outputs, 150 ns.)
PAL	Monolithic Memories PAL10H8 (10 inputs, 8 outputs, 16 terms)
Registered PAL	Monolithic Memories PAL16X4 (8 inputs, 8 outputs, 4 registered outputs, 64 terms)
ROM: EROM	Signetics SN 74S271/371 (256x8 with 45 ns access time)
PROM	Signetics SN 74S271/371 (512x8 with 55 ns access time)
Registered PROM	AMD Am27S26/27 (512x8 with 20 ns Cp to output time-prelim)

Table 11.1. Examples of
Microprogrammable Devices.

11.5.1 Programmable Multiplexers

One of the programmable multiplexers is the Raytheon 29693 PMUX listed in Table 11.1. This device has 10 inputs, all inverted, and 4 inverted outputs. Logically it is equivalent to 4, 1-of-8 multiplexers with 10-input OR gates as the inputs. The multiple-output problem presented earlier is of the type suitable for a PMUX implementation because: (1) The MUX's have common select lines; (2) the number of input variables is less than 10. Figure 11.7 is very nearly a program for the 29693 PMUX. The only change required is to use $X_3 + \bar{X}_3$ as the input where a connection to V_{CC} is shown. Documentation of this device is one to four Marquand maps (one for each output used) drawn with select variables as column indices and input variables as row indices. Note that an external inverter for X_3 would be required. Note also that this example under-utilizes the PMUX's ability and is for demonstration only.

11.5.2 Programmed Logic Arrays

A Programmable Logic Array (PLA) is an LSI implementation of the classic digital net, a sum-of-products form for positive logic. It is general-purpose and user-definable within limits. Both the variable composition of the individual product terms and the assignment of the terms to the different outputs are specified by the designer.

PLA's are characterized by the following attributes:

(1) The number of inputs

(2) Buffered or unbuffered inputs

(3) The number of minterms or product terms

(4) The number of outputs

    (5)   The characteristics of the outputs:

        a.   TTL

        b.   Open-collector (OC)

        c.   Tri-state

    (6)   Programmable output inversion

    (7)   Access time

There are two basic types of PLA's, determined by how they are programmed. A factory- or mask-programmed PLA is typically larger with 96 minterms. Factory programming uses the metal mask technology.

A field-programmable PLA (FPLA) typically has 48 minterms and may use one of three programming technologies:

    (a)   Nichrome fuse

    (b)   Polysilicon fuse

    (c)   Avalanche induced migration

Given a PLA with the following characteristics:

    (1)   16 inputs, either true or inverted, to any AND gate,

    (2)   8 outputs, either true or inverted,

    (3)   48 product terms to any OR gate,

    (4)   Typical access time of 50 ns,

    (5)   Power dissipation of 620 mW (typical),

    (6)   28-pin DIP (approximate space of 3 16-pin DIP's);

the six-variable functions shown in Figure 11.10a are to be implemented. These functions are not necessarily minimal. These are, using the available part above, 6 inputs required, 14 product terms, of which 3 are immediately seen to be identical, and 4 outputs. If the equations given had not been in a sum-of-products form, expansion would have to be performed to obtain this format as each output is a sum-of-products function. Internally the PLA is an AND-OR net with an inverter programmable at the output if desired.

$$y_0 = x_0 \bar{x}_1 x_2 + \bar{x}_1 \bar{x}_3 + x_1 x_2 \bar{x}_3 x_4 \bar{x}_5 + \bar{x}_4 \bar{x}_5$$

$$y_1 = x_0 \bar{x}_1 x_2 x_5 + \bar{x}_1 \bar{x}_3 + \bar{x}_1 x_4 x_5$$

$$y_2 = x_2 \bar{x}_3 x_5 + \bar{x}_0 x_1 \bar{x}_2 x_3 \bar{x}_4 x_5 + \bar{x}_1 x_2 \bar{x}_5$$

$$y_3 = x_0 x_2 x_5 + \bar{x}_1 x_4 x_5 + \bar{x}_1 \bar{x}_3 + x_0 x_2 \bar{x}_5$$

must be in

$\Sigma\Pi$ form

a. The equations.

4 functions   PLA chosen has 4 outputs,

6 input variables   6 inputs,

14 terms   13 product terms

$x_0 x_1 x_2 x_3 x_4 x_5$	$y_0 y_1 y_2 y_3$
1 0 1 x x x	1 0 0 0
x 0 x 0 x x	1 0 0 0
x 1 1 0 1 0	1 0 0 0
x x x x 1 1	1 0 0 0
1 0 1 x x 1	0 1 0 0
x 0 x 0 x x	0 1 0 1
x 0 x x 1 1	0 1 0 0
x x 1 0 x 1	0 0 1 0
0 1 0 1 0 1	0 0 1 0
x 0 1 x x 0	0 0 1 0
1 x 1 x x 1	0 0 0 1
x 0 x x 1 1	0 0 0 1
1 x 1 x x 0	0 0 0 1

← share one term, all that is required to make the functions fit the PLA (in this case).

b. The PLA program.

Figure 11.10. Designing with a small PLA.

For the functions shown in Figure 11.10a, a single PLA may be programmed as shown in Figure 11.10b. No minimization needs to be performed.

Svoboda's six-variable problem could easily be implemented with one PLA-type device. There are 20 product terms required, less than one-half the number available in the smallest of these devices. The product terms can be trivially taken from the Marquand map of the function without minimization. The tradeoffs are increased power consumption for less board space, a simpler design effort, and reduced testability.

(Note that under-utilization of the LSI devices is quite common and is cost-effective for commercial applications.)

Where the product terms exceed the number (48, 96) available by a small amount, some minimization may be performed to reduce the functions to fit the boundaries. Svoboda's multiple-output minimization is suitable for these cases. Where there are enough product terms to warrant it (double or triple the number available), a second PLA may be parallelled with the first to produce the added capability. All input and output lines would be common.

Where the number of outputs exceeds that available (8) and a simple added SSI or MUX unit is not sufficient, PLA's may be parallelled where the input lines are tied common and the output lines are separate.

Combinations of these two schemes may be used to implement odd-sized problems. Note that open-collector outputs and low-level logic is required for any of these cases.

An important point with PLA's is that they are pure digital internally. In cases where the design is sensitive to signal line glitches, PLA's may be preferable to naked ROM's and should be considered.

It should be noted that, while an 18-input ROM would require 256K internal cells and is several years away by present technology, PLA's can provide a subset of the same logical space today.

11.5.3 Programmable Array Logic

PAL's are also referred to as programmable gate arrays. The PAL product line of Monolithic Memories is characterized by 8 to 16 inputs, available internally in true and complemented form, and 2 to 8 outputs, with a varying number of product terms. The low end of the product line is the PAL10H8 with 10 input variables, 8 outputs, and two product terms per output.

A PAL is different from a PLA in that the number of terms per output is fixed, imposing more constraints on the designer. Some PAL's are available with feedback and with registered output and feedback. While a PLA replaces SSI combinational logic, PAL's exist which can replace blocks of SSI sequential logic. The PAL16R6, for example, has 8 inputs, 8 outputs of which 6 are registered, with all outputs fed back. The interested designer should refer to the Monolithic Memories <u>PAL</u> <u>Handbook</u>.

The six-variable problem of Svoboda requires 20 product terms, 6 inputs and 1 output. By using a small amount of minimization to reduce the number of product terms, a PAL could be utilized. Figure 11.11 shows the program connections to construct the solution using a PAL10H8 and an OR gate. The terms programmed are from the multiplexer implementation of the problem shown in Figure 11.8. Column minimization was applied to the Marquand map of the function in both cases. The PAL replaces two 1-of-8 MUX's and five NOR gates. The OR gate is in exchange for the EXOR of Figure 11.8.

Logic Diagram PAL10H8

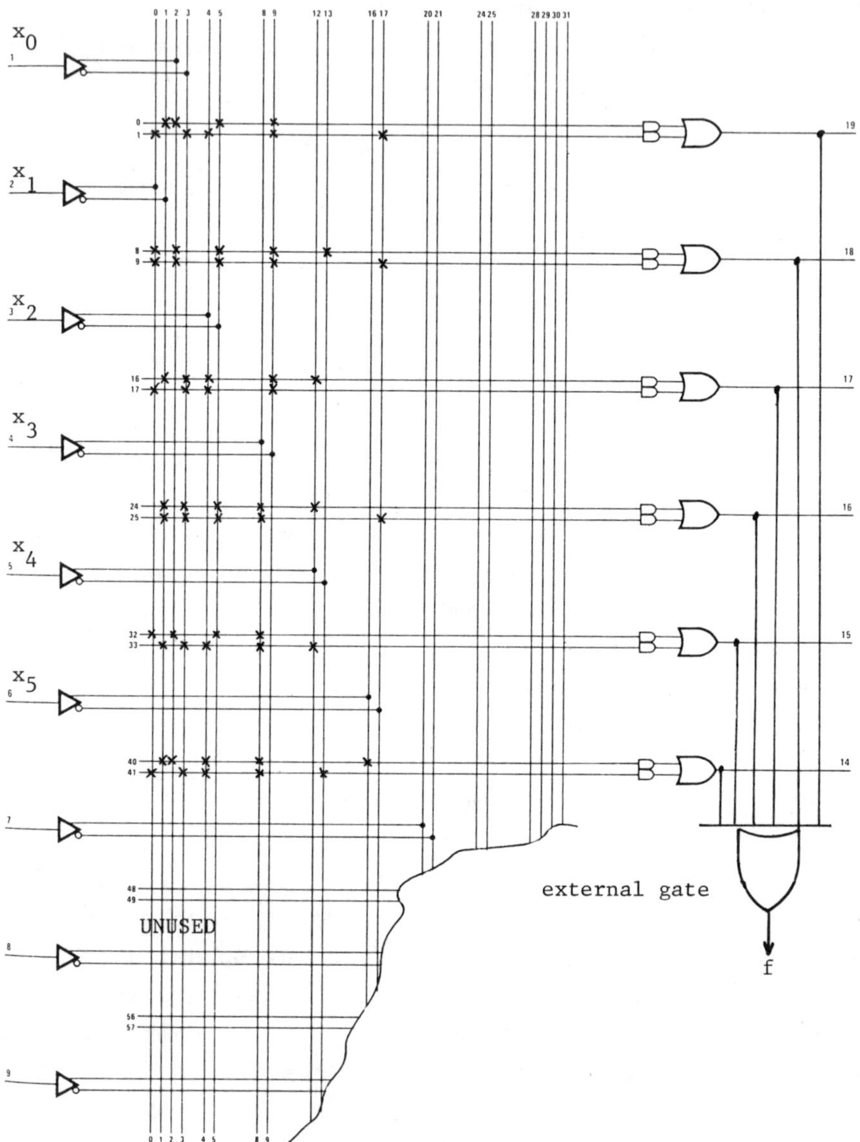

Figure 11.11.   Implementation of the six-variable problem with a PAL.

11.5.4 Design Example

Using Baucham's modified Hamming code for high-speed single-error correct, double- (and some triple-) error detection (Figure 11.12) which is for use with 16-bit-word memories, most of the parity equations of the code can be implemented via simple SSI and MSI devices. The parity equations for the generation of check bits may be implemented with six 74S280 parity generators (Figure 11.13a).

Decode logic is also similarly based on parity generators to provide the syndrome check bits (Figure 11.13b). The overall syndrome bit S_D is computed using all data bits and all check bits. Using P_e for even parity, P_o for odd parity:

$$S_D = P_e\ (C_1,\ C_2,\ C_3,\ C_4,\ C_5,\ C_D)$$
$$+ P_o\ (D_0,\ \ldots,\ D_7) + P_o\ (D_8,\ \ldots,\ D_{15})$$

However, Syndrome bit 5 is found from
$$S_5 = P_o\ (D_8,\ \ldots,\ D_{15},\ C_5)$$
To keep part counts down, redefine

$$S_D = S_5 + P_o\ (D_0,\ \ldots,\ D_7)$$
$$+ P_o\ (C_1,\ C_2,\ C_3,\ C_4,\ C_D)$$

Then use a multiplexer to generate S_D (see Figure 11.14).

Given the syndrome bits, a decode operation is necessary to determine the error status as: (1) No error has occurred; (2) one and, therefore, a correctable error has occurred; or (3) more than one error has occurred (uncorrectable). The decode matrix is specified via a Marquand map in Figure 11.15. The status signals may be generated using two 1-of-16 multiplexers (Figure 11.16). Demultiplexers can decode the individual bit to be corrected and correction accomplished by inverters and XOR gates.

$$
\begin{bmatrix}
1\;1\;1\;1\;1\;1\;1\;1\;1\;1\;1\;1\;1\;1\;1\;1\;1\;1\;1\;1\;1\;1 \\
1\;1\;0\;0\;0\;0\;0\;0\;1\;1\;1\;1\;1\;1\;0\;0\;0\;1\;0\;0\;0\;0 \\
0\;0\;1\;1\;1\;0\;0\;0\;1\;1\;1\;0\;0\;0\;1\;0\;0\;0\;1\;0\;0\;0 \\
1\;0\;1\;0\;0\;1\;1\;0\;1\;0\;0\;1\;1\;0\;0\;1\;0\;0\;0\;1\;0\;0 \\
0\;0\;0\;1\;0\;1\;0\;1\;0\;1\;0\;1\;0\;1\;1\;1\;0\;0\;0\;0\;1\;0 \\
0\;1\;0\;0\;1\;0\;1\;1\;0\;0\;1\;0\;1\;0\;1\;1\;0\;0\;0\;0\;0\;1
\end{bmatrix} = H
$$

$$
I_{15}\,I_{14}\,I_{13}\,I_{12}\,I_{11}\,I_{10}\,I_9\,I_8\,I_7\,I_6\,I_5\,I_4\,I_3\,I_2\,I_1\,I_0\,\bar{C}_5\,C_4\,\bar{C}_3\,C_2\,\bar{C}_1\,\bar{C}_D
$$

C_D, C_1, C_3, C_5 are all complements   (odd parity)

C_2, C_4 are even parity sums

Syndrome bits are computed by:

$$S_i = C_i^1 + C_i$$

C_i^1 is computed upon receipt of the data

C_i is computed and stored and transmitted with the data

$$S_D = P_e\,(\,C_1,\ldots,C_5,C_D\,) + P_o\,(\,D_0,\ldots,D_7\,) + P_o\,(\,D_8,\ldots,D_{15}\,)$$

P_e = even parity

P_o = odd parity sum

C_i = check bit

D_i = data or information bit

Figure 11.12.   Basham's modified Hamming code.

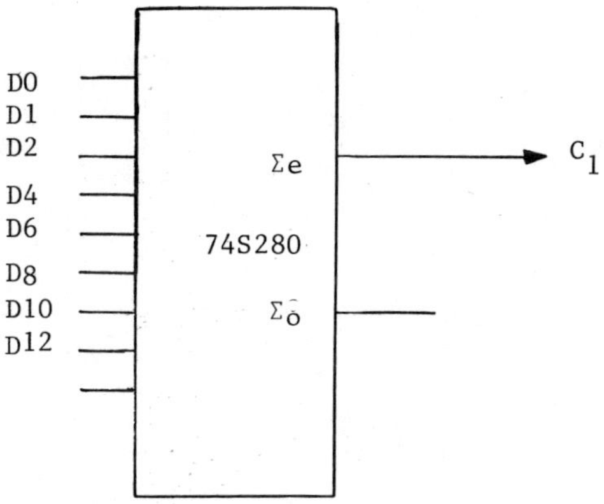

a.   Odd parity check bit generation.

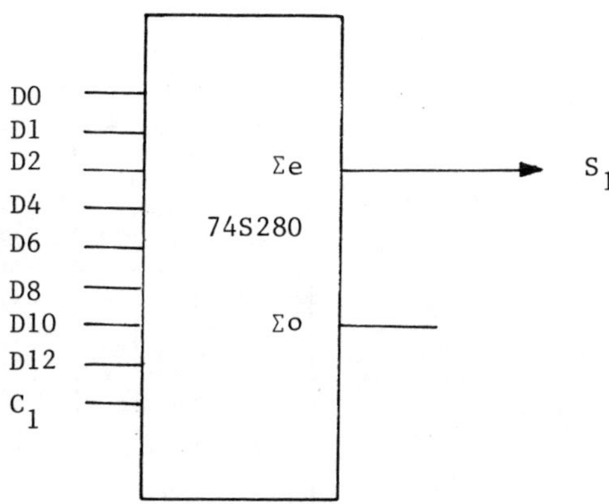

b.   Syndrome bit generation.

Figure 11.13.   Parity and check bit generation.

$$C_p = P_o \ (\ C_1, C_2, C_3, C_4, C_D \ )$$

$$P_1 = P_o \ (\ D_0, D_1, \ldots, D_7 \ )$$

$$S_5 = \text{syndrome bit 5}$$

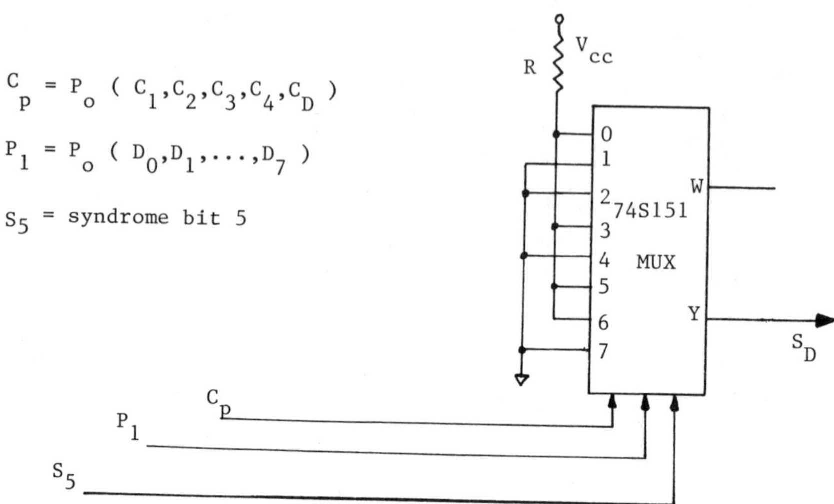

Figure 11.14. Syndrome bit generation using a multiplexer.

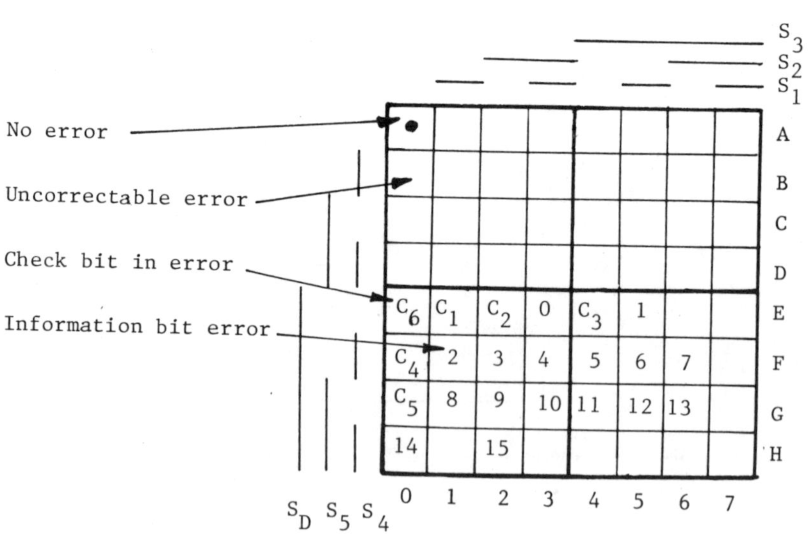

Figure 11.15. Syndrome bit decode.

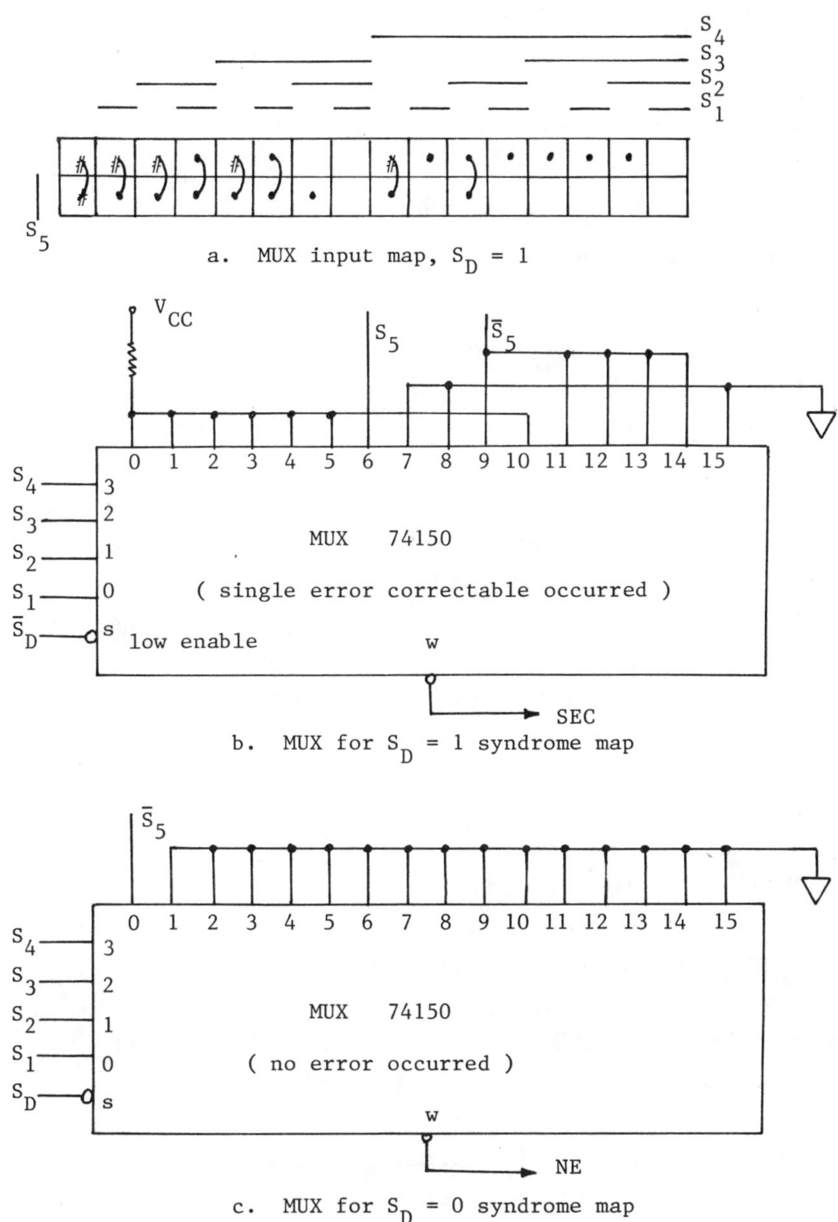

a.   MUX input map, $S_D = 1$

b.   MUX for $S_D = 1$ syndrome map

c.   MUX for $S_D = 0$ syndrome map

Figure 11.16.   Control signal generation.

The syndrome bit decode could also be implemented using a single PAL. Here, Marquand maps can provide the necessary documentation and can be used as a tool for minimization. A PAL such as the PAL16L8 can be used to replace the two 1-of-16 MUX chips and an AND gate. The coding for the PAL16L8 is shown in Figure 11.17.

11.5.5 Read-Only Memories

Read-Only Memories or ROM's (Figure 11.18) contain a bit position for every combination of input variable validities for each of its multiple outputs. A ROM may be thought of as containing the information from m Marquand maps where m is the number of outputs. A ROM is characterized by: (1) Its programmability (field, PROM, or factory, ROM); (2) its reprogrammability (eraseable, EPROM, or not); (3) whether it has output latches or registers (registered-PROM), (4) its size, expressed as the number of addresses (2^n where n is the number of inputs); (5) the number of outputs; (6) whether it is all "1"s or all "0"s initially; (7) its speed (access time if unregistered, set-up and clock-to-output if registered); (8) its output (open-collector or tri-state); and (9) its enable structure.

ROM family elements (EPROM, PROM, etc.) are useful, for example, as decoders and as sequential controllers. Using a ROM represents a shift from high-speed, custom, hardwired SSI logic to possibly slower, flexible, LSI firmware design. A ROM plus a sequencer such as the AM2910 represents a structured approach to sequential design. It may be said the microprogramming is to hardware design what structured programming is to software design. There are parallel tradeoffs which must be considered in each case.

Unlike the PGA, PAL, PLA, and PMUX devices, a ROM may be erased (EPROM). Both PAL's and ROM's have deve-

Logic Diagram PAL16L8

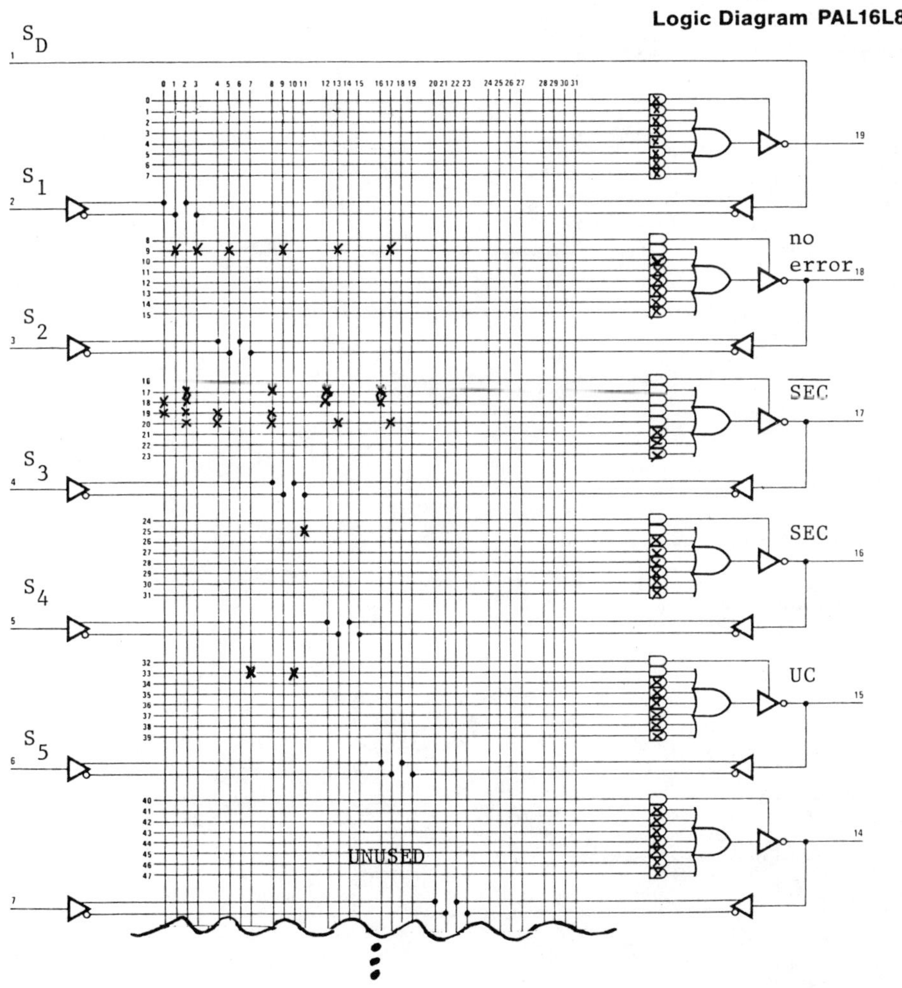

Figure 11.17.   Programmed PAL for control signal generation.

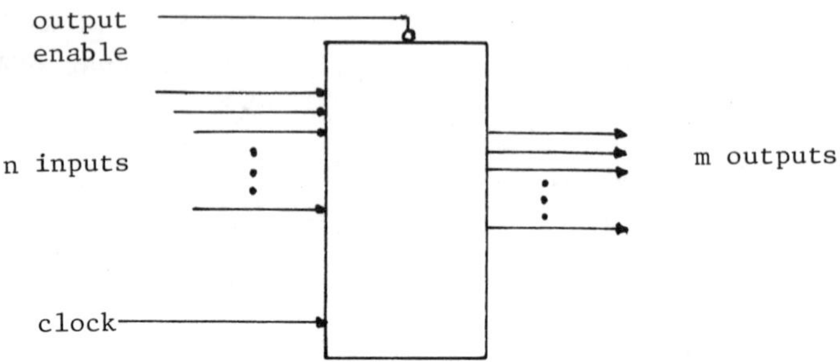

n inputs address 2^n words of m bits per word

Figure 11.18.   ROM logic block.

DEC equiv address:	y outputs		
	0	1	2
0	0	0	1
1	1	1	1
2	0	1	1
3	0	1	1
4	1	0	0
5	1	0	1
6	1	0	0
7	1	0	0
8	1	0	1
9	0	0	0
10	1	0	1
11	0	0	0
12	1	1	1
13	0	0	0
14	1	0	0
15	0	0	1

program

1 16-pin DIP, 16 connections, 575 mW,
40 ns access (worst case)

Figure 11.19.   Multiple output problem implemented in ROM.

lopment system assemblers available to assist in their programming. A ROM is documented by the assembly program which prepares its tape or by a manually-prepared mnemonic program listing. (Designing with bit-slice architecture and microprogramming in general are not covered in this text.) Where the circuit is sensitive to glitches in signal lines, registered PROM's or external latches are preferable to "naked" ROM's.

Figure 11.19 demonstrates the multiple-output problem implemented in a 32 x 8 bipolar PROM. Svoboda's six-variable example is given in Figure 11.20 using a 256 x 4 bipolar PROM. In any design, the fact that there are unused areas in the PROM is not a problem. Unused areas allow for minor modifications without major redesign or for patches to the "microprogram".

11.5.6 Sequential Design Example

To demonstrate microprogramming as a sequential design tool, a traffic light controller is included here. Note that this is not a valid application for the AM2910, but is sufficient to demonstrate the use of this device in addition to being a problem of manageable size for discussion.

A sequencing schemata is given in Figure 11.21. The starting point is main-street-green (MG). During main-street-yellow (MY) a test is made to determine if a protected left turn (MLT) is desired. If it is, main-left-turn-green and main-left-turn-yellow are cycled. Next, the side street turns green (SG) for its time period. During side street yellow (SY), tests are made for manual override, where the lights will all flash red (accident control); for night, where the lights will all flash red (STOP) except for the main street lights which will flash yellow (CAUTION); and for side protected left turn, in this priority.

Program in address - word format

0 0000	16 0000	32 0000	48 0000
1 0001	17 0001	33 0000	49 0000
2 0000	18 0001	34 0000	50 0000
3 0000	19 0001	35 0001	51 0000
4 0000	20 0001	36 0000	52 0001
5 0000	21 0000	37 0000	53 0000
6 0000	22 0001	38 0001	54 0000
7 0000	23 0000	39 0000	55 0000
8 0001	24 0001	40 0000	56 0001
9 0000	25 0000	41 0000	57 0000
10 0000	26 0000	42 0000	58 0000
11 0000	27 0001	43 0001	59 0001
12 0000	28 0000	44 0000	60 0001
13 0000	29 0000	45 0000	61 0001
14 0001	30 0000	46 0001	62 0000
15 0000	31 0000	47 0000	63 0001

Since this PROM is all 0 before programming, all # are taken as 0.

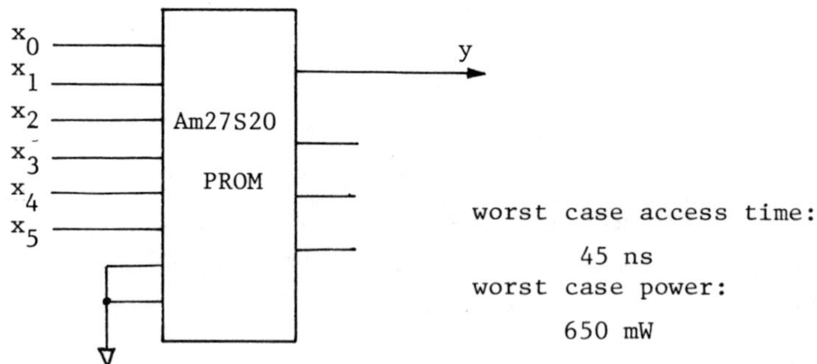

Figure 11.20.   Svoboda's six-variable example done with a PROM.

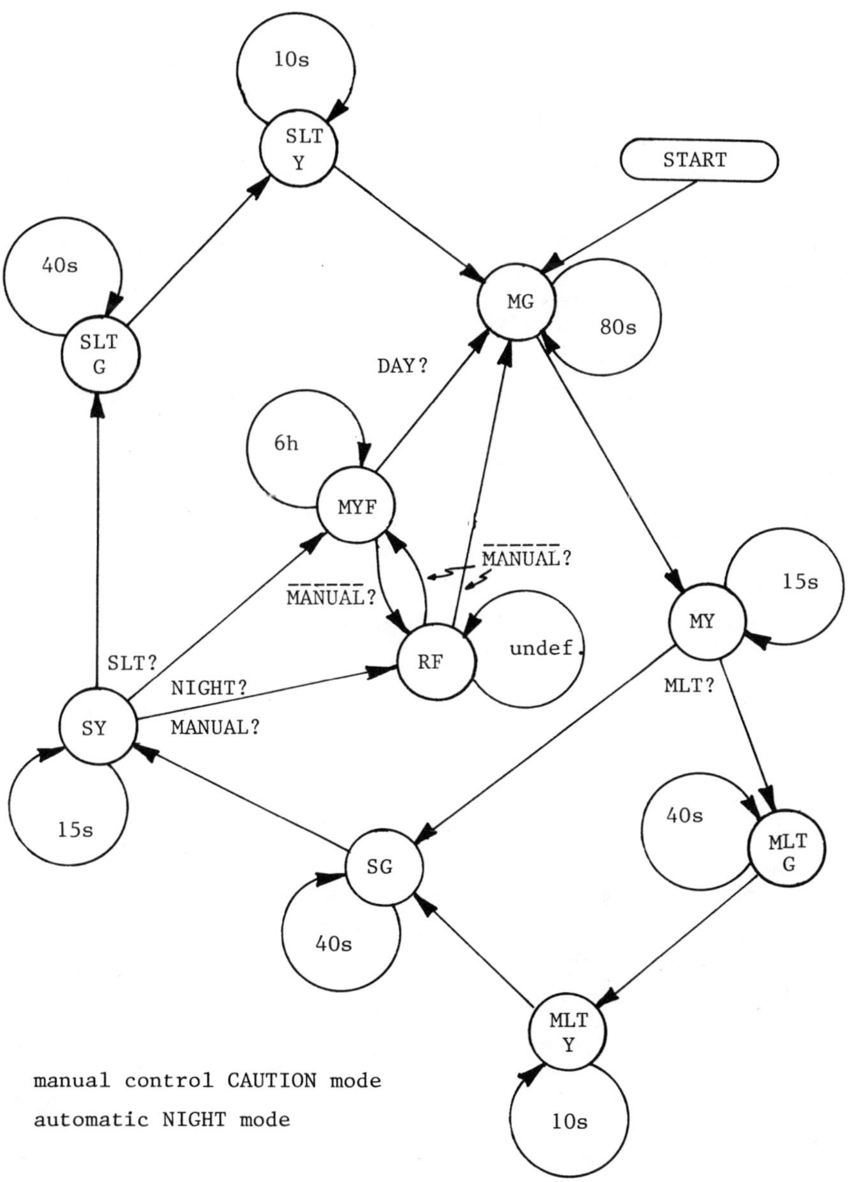

manual control CAUTION mode
automatic NIGHT mode

Figure 11.21. Traffic light controller.

If MANUAL is present, set all lights red and loop here until $\overline{\text{MANUAL}}$ occurs. If $\overline{\text{MANUAL}}$ occurs, return to normal cycle (or go to night mode if NIGHT is present). If NIGHT is present, set the lights and loop until DAY (= $\overline{\text{NIGHT}}$) or MANUAL occurs. The times are in seconds and a 5-second clock is assumed.

The lights are controlled by three signal lines each to provide signals RED, YELLOW, GREEN, RED FLASH and YEL-LOW FLASH. The state sequenced paths for the light controls are given in Figure 11.22 and are considered a fixed constraint for the purposes of the problem. (That may seem arbitrary, but remember that this is for demonstration.)

The control portion of the system is shown in Figure 11.23 and uses one AM2910 sequencer, 3 32 x 8 PROM's (registers are not necessary here; the pipeline is shown only to demonstrate output enable), and one MUX. MLT, $\overline{\text{SLT}}$, NIGHT, MANUAL and START are assumed to be available. A microprogram for the controller is given in Figure 11.24 using mnemonics rather than bit patterns. (A System 29 development system can be used to generate the bit patterns for the PROM's.) The AM2910 has 16 instructions which facilitate programming. A summary of those used are given in Table 11.2. A direct similarity can be drawn to FORTRAN-type or assembly-level languages.

Name	Actual Pattern	(Main lights)	Name	Actual Pattern	(Side lights)
RO	000		Ro	000	
R2	010	red	R2	010	red
R4	100		R4	100	
Y	010	yellow	Y	010	yellow
G	011	green	G	011	green
YF	101	yellow flash	RF	101	red flash
RF	110	red flash			

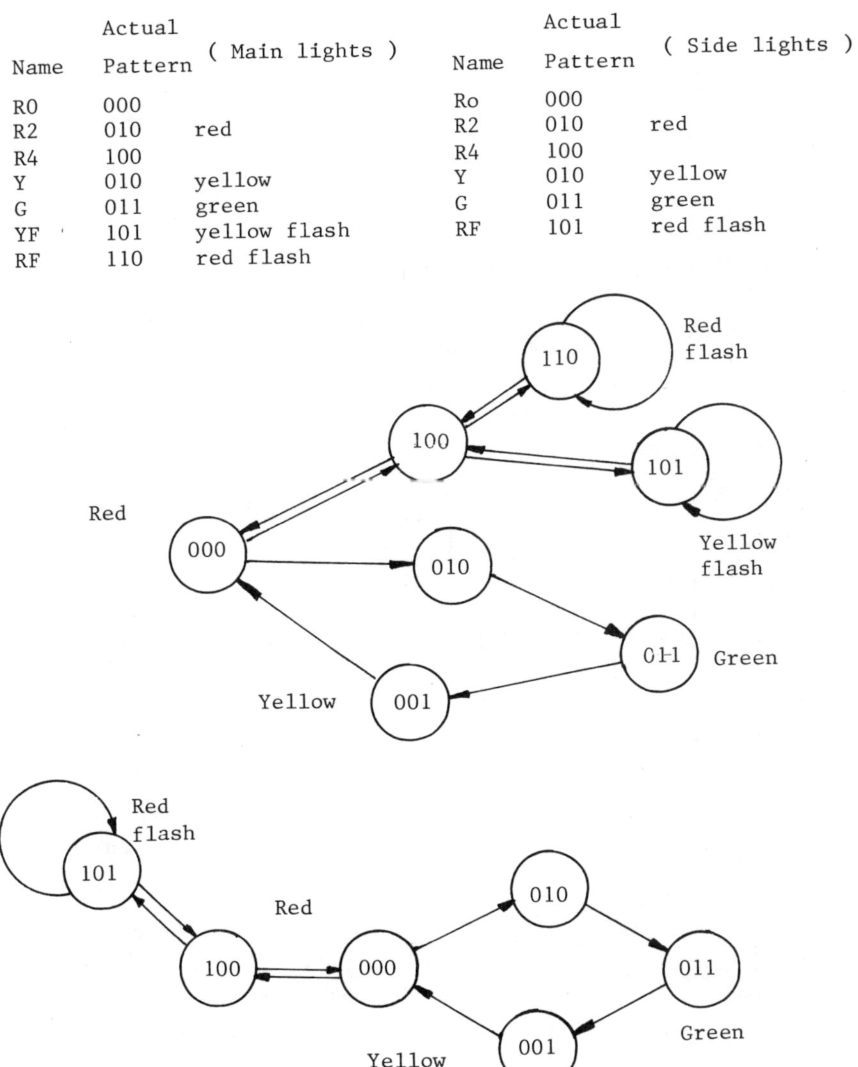

Figure 11.22. Light control signal state sequencing.

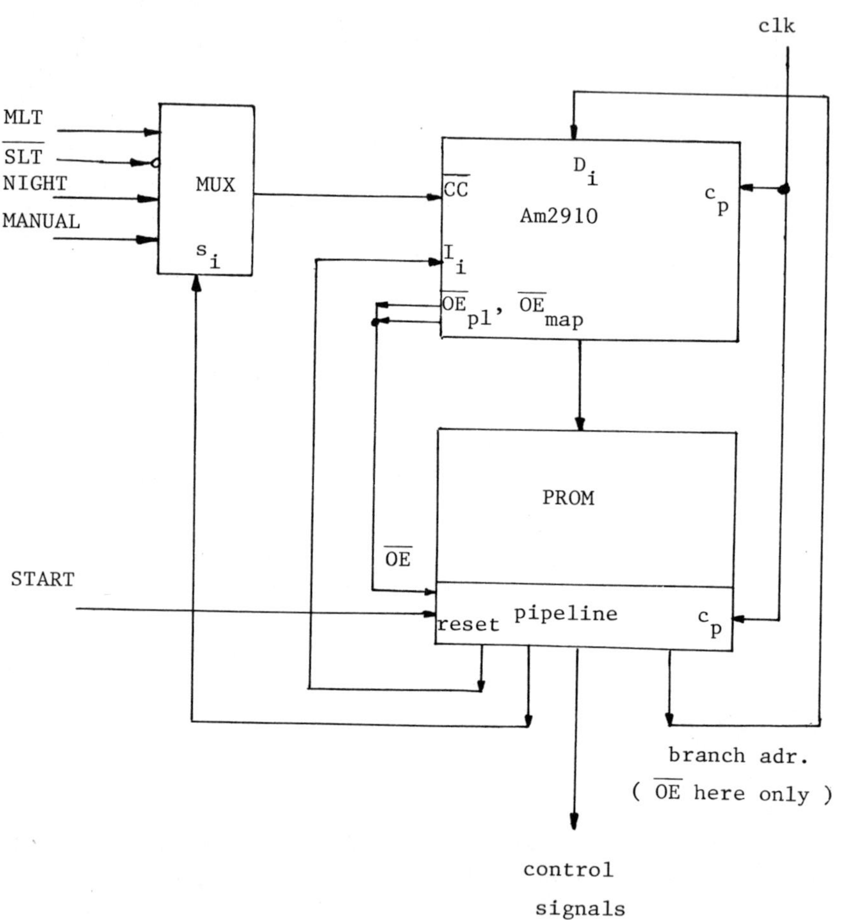

Figure 11.23.   Traffic light control design with the Am2910.

DEC ADR	LABEL	2910 INSTR	MUX TEST	COUNTER or BRANCH	S	MLT	SLT	M	COMMENT
		SEQUENCE CONTROL			LIGHT CONTROL				
0	MAIN	LDCT		15	RO	RO	RO	R2	LOAD CNTR
1	MLP	RPCT		MLP	RO	RO	RO	G	LOOP PL-MAIN
2		CONT			RO	RO	RO	Y	
3		CJP	MLT ?	MLT	RO	RO	RO	Y	MAIN LT TEST
4		LDCT		7	R2	RO	RO	Y	
5	SIDE	RPCT		SIDE	G	RO	RO	RO	LOOP PL- SIDE
6		CJP	MANUAL?	MANUAL	Y	RO	RO	RO	MANUAL OVERRIDE?
7		CJP	NIGHT?	NIGHT	Y	RO	RO	RO	NIGHT TEST?
8		CJP	SLT?	MAIN	Y	RO	RO	RO	SIDE LT TEST
9	SLT	LDCT		7	RO	RO	R2	RO	SIDE LT LOOP
10	SLTLP	RPCT		SLTLP	RO	RO	G	RO	LOOP PL
11		CONT			RO	RO	Y	RO	
12		JMAP		MAIN	RO	RO	Y	RO	MAP TIED TO PL OE
13	MLT	LDCT		7	RO	R2	RO	Y	MAIN LT LOOP
14	MLTLP	RPCT		MLTLP	RO	G	RO	RO	LOOP PL
15		LDCT		7	RO	Y	RO	RO	
16		JMAP		SIDE	R2	Y	RO	RO	
17	NIGHT	LDCT		16	Y	R4	R4	R4	altered
18	NITLP	RPCT		NITLP	RF	RF	RF	YF	sequence
19		LDCT		16	RF	RF	RF	YF	
20		CJP	MANUAL?	FIX	RF	RF	RF	YF	
21	NTEST	CJP	NIGHT?	NITLP	RF	RF	RF	YF	CONTINUE NITEMODE
22		CONT			R4	R4	R4	R4	
23		JMAP		0	RO	RO	RO	RO	
24	MANUAL	CONT			Y	RO	RO	RO	altered
25		LDCT		16	Y	R4	R4	R4	sequence
26	ENTRY	RPCT			RF	RF	RF	RF	LOOP PL
27		LDCT		16	RF	RF	RF	RF	
28		CJP	MANUAL?	ENTRY	RF	RF	RF	RF	
29		JMAP		NTEST	RF	RF	RF	RF	←share jump
30	FIX	JMAP		ENTRY	RF	RF	RF	R4	
31		JZ		MAIN	RO	RO	RO	RO	

Figure 11.24. Microprogram for the controller.

Instruction	Function
CONT	continue, address = address + 1
LDCT	load counter and continue
RPCT	repeat starting at given address until counter = 0 (DO LOOP)
JMAP	GO TO branch address
CJP	Conditional jump (IF – THEN)
JZ	Initialize jump zero

Table 11.2.   Table of Am2910 instructions used.

Chapter 12
Fault Detection Techniques

12.1 FAULTS

12.1.1 Fault Definition

In any circuit composed of logic gates there is the
possibility of the occurrence of a fault. A fault is
defined to have occurred when any circuit variable
assumes a value (1, 0, or X) which differs from that
expected, that is, which violates the original circuit
equations.

12.1.2 Masking a Fault

The presence of an internal or input fault may not
be observable at the circuit output in which case the
fault is considered to be masked. A single fault may be
masked as the result of: (1) Reconvergent fanout where
unequal parity changes have occurred; (2) circuit redundancy; and (3) the previous occurrence of an undetectable
fault. Masked faults are undetectable by their definition
since the observed circuit behavior is correct. However,
the occurrence of a second fault may uncover a previously-
undetectable fault. To be complete, the test set must
include tests for this case.

Figure 12.1 presents a redundant circuit and its
Marquand map. The circuit implements three terms to
cover the eight points on the map when the two terms,
$\bar{x}_3 x_0$ and $x_3 x_1$, are sufficient.

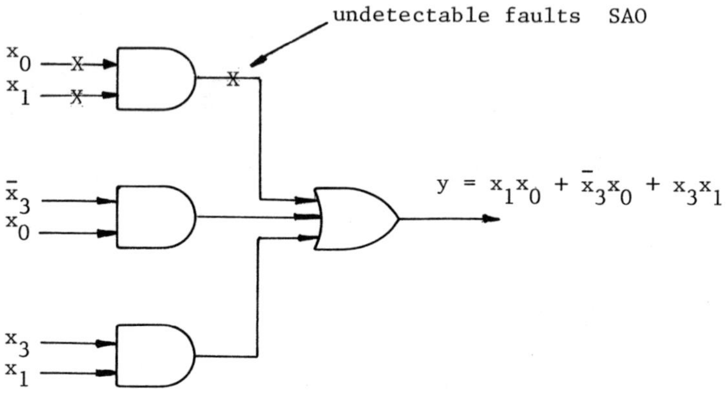

a.   The circuit

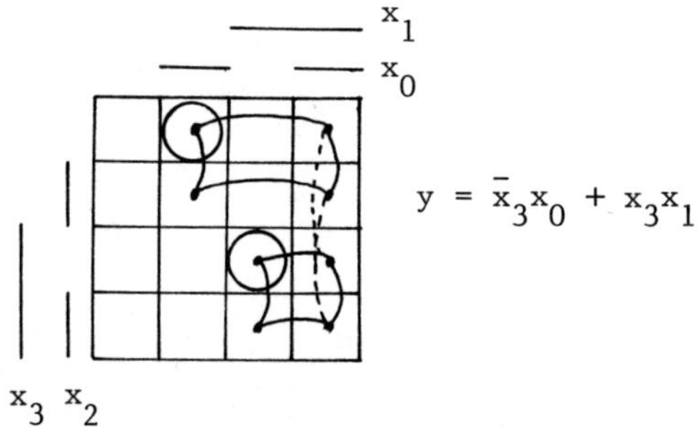

b.   Marquand map

Figure 12.1.   Redundant circuit.

FAULT DETECTION TECHNIQUES

12.1.3 Fault Types

Faults may be indeterminate in value (suspended between logical '1' and logical '0') or determinate in value (exhibiting a '0' or a '1').

Faults may be transient (intermittent, time varying), in which case they are elusive and difficult to detect. Faults may be permanent (considered "hard" or "solid"), in which case they are easy to detect if they are not masked and if a proper test is used. Faults may be multiple in occurrence, which has always been considered a rare event. (With the higher circuit densities, this event has increased somewhat in probability.) Faults may occur singly, which is considered to be the most likely event. Further, multiple faults can occur in such a manner that there is an equivalent single fault for them. A test which detects the presence of this equivalent fault will be sufficient to detect the presence of faults. It should be noted that fault <u>identification</u> is not possible.

12.1.4 Fault Equivalencies

There are several equivalencies that exist which are useful in fault detection and which make fault location considerably more difficult. Some of these equivalencies are: (1) One or more inputs to an OR gate stuck at '1' (SA1) is equivalent to the output of the OR gate SA1; (2) one or more inputs to an AND gate stuck at '0' (SA0) is equivalent to the output of the AND gate SA0; (3) all inputs to an OR gate SA0 is equivalent to the output of the OR gate SA0; (4) all inputs to an AND gate SA1 is equivalent to the output of the AND gate SA1; (5) failures on both the inputs (one or more failures) and the output of a gate will propagate the gate output failure, masking the input faulting; and (6) any gate output fault has, as an equivalent, a single gate input fault (not necessarily an input to that gate) or multiple input faults. However, any gate input fault does not necessarily have an equivalent gate output fault.

12.1.5 The Problem

The most common fault for current technology (such as DRL, DTL, RTL, and TTL) is the single, permanent, stuck-at fault where one of the following has occurred: (1) A gate output is stuck at logical '0' (SAO); (2) a gate output is stuck at logical '1' (SA1); (3) any single gate input line is stuck at '0'; or (4) any single gate input line is stuck at '1'. The single fault assumption is not proper for the initial circuit checkout.

Component failures which alter or affect voltage levels, current levels, pulse widths, or circuit timing, but which do not alter or affect the logical function realized by the circuit, will not be considered here. These qualitative failures are presumed detected during initial electrical parameter testing.

To be complete, a test set must be able to detect any single detectable fault. It should also inlcude tests for multiple faults, where such faults are not covered by equivalent single faults. Further, the test set should include tests for faults which become detectable when another undetectable fault occurs (this is a special type of multiple fault).

The circuit of Figure 12.2 has seven lines and either a SAO or SA1 fault may occur on any line. There are, threfore, fourteen single faults possible for this case. There are, by computation, 84 possible double faults which may occur. All 84 double faults are "covered" by the fourteen single faults. Further, of the fourteen single faults, not all are distinct. As an example, a test for x_0 SAO also tests x_4 SAO and x_6 SAO. This allows a test set to be derived which is smaller

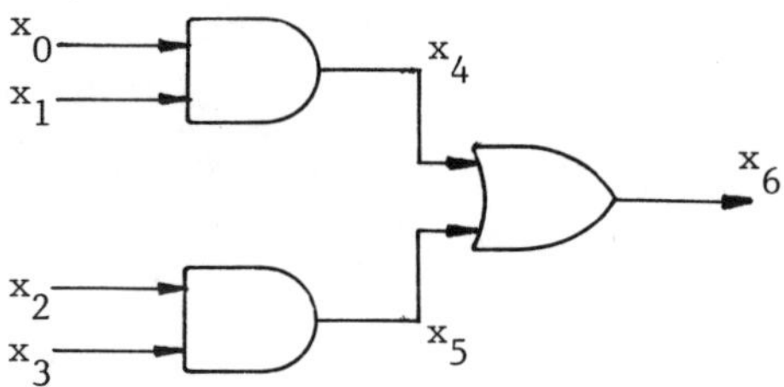

a.   TNC labeling for a two-level AND-OR circuit.

Equation:

$$x_6 = x_4 + x_5 = x_0 x_1 + x_2 x_3$$

Number of fault locations: 7

Number of possible single faults: $\binom{2}{1}\binom{7}{1} = 14$

Number of double fault combinations:

$$\binom{2}{1}\binom{2}{1}\binom{7}{2} = 84$$

Figure 12.2.   Sample circuit used in comparing methods.

than the exhaustive test set (2^4 = 16 tests for this
case) and often smaller than that produced with the "one-
test-per-fault" approach.

If it is assumed that one or more permanent stuck-
at faults (SA1 or SA0), or the equivalent, has occurred,
then the problem is to construct a complete and minimal
test set such that this fault condition is detected, provided that masking has not covered the effects.

12.2   THE TEST SEQUENCE

The original paper on Boolean differences by this
author, written as a class exercise in 1971, began the
search to find a method of fault detection which would
be produceable by the Boolean Analyzer. In the course
of studying the fault detection problem, a number of
papers, several of which have been used as references,
were examined and the various methods presented cata-
logued by their approach to the problem.

With the discovery of certain properties of the edge
structure of the Existence Function, which allows the
formation of links, and the first hypothesis for sequence
construction, most of the examples from those papers were
reexamined. In each case, both the original procedure
and the Test Sequence procedure were performed and the
results were compared. For combinational circuits, the
limit of this research, the results in all cases were
favorable.

12.2.1   Deriving the Existence Function

12.2.1.1   The Equations. The first step in developing
a Test Sequence is the derivation of the equations of a
given circuit configuration. Equations are first derived
for each of the logic gates, using the circuit primary
inputs as known variables, and all intermediate and primary outputs as unknown variables.

FAULT DETECTION TECHNIQUES

A primary input is an input connected to an external
source. A primary output is an output goint to an external "sink" or connection. All other lines are to be considered as internal or intermediate lines and their
labels are handled as intermediate variables. For the
purposes of this paper, a logical gate will be cons-
trained to be an SSI level gate (AND, OR, NAND, NOR, or
INV).

As an example of this first type of equation derivation, there are three equations for the first circuit,
Circuit 1, of Figure 12.3. They are:

$$x_6 = x_3 x_4 \qquad x_5 = x_1 x_2 \qquad x_7 = x_5 + x_6$$

Labelling for this case has begun with x_1.

For the case where there is fanout of primary input
lines, the first line is labeled and treated as a known
variable while the remaining fanout lines are labeled
and treated as intermediate variables. Equations must
be added to the equation set which equate the first line
with the remaining fanout lines.

For example, the second circuit, Circuit 2, of
Figure 12.3 has the two equations:

$$x_5 = x_1 \qquad x_8 = x_7$$

added to the listed equations for the input/output relationships of the circuit.

Internal fanout and fanin are handled in a similar
manner. Additional examples are presented later
to demonstrate the procedure.

12.2.1.2 Required Labelling. There is a labelling convention which has been used in the above equations and
which has been given the name TNC (Terminal Numbering
Convention) in previous papers. Under this convention,
the primary input lines are labeled with the lowest

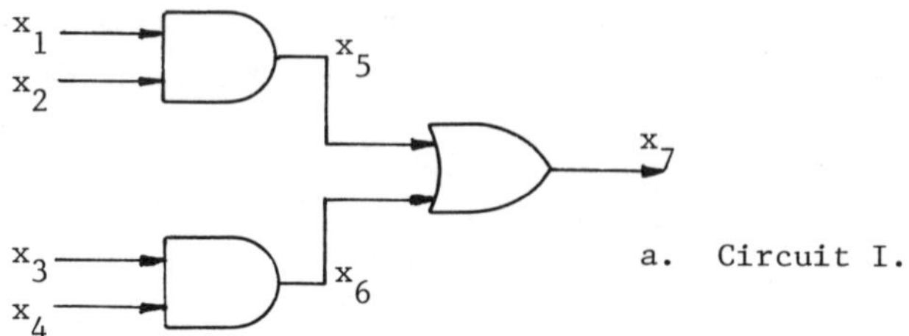

a. Circuit I.

Equations for the gates:

$$x_5 = x_1 x_2$$
$$x_6 = x_3 x_4$$
$$x_7 = x_5 + x_6$$

Equations for fanout: none

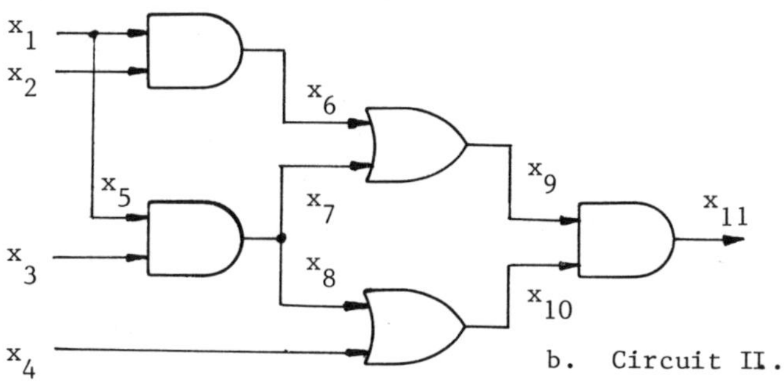

b. Circuit II.

Equations for the gates:

$$x_6 = x_1 x_2 \qquad x_9 = x_6 + x_7$$
$$x_7 = x_3 x_5 \qquad x_{10} = x_4 + x_8$$
$$x_{11} = x_9 x_{10}$$

Equations for fanout: $x_5 = x_1 \qquad x_8 = x_7$

Figure 12.3. Examples of labeling; equations.

indexed variables (x_0, x_1 are the two most common starting labels). Each gate input and output receives a
distinct label with the convention that, for any gate,
the indices of the labels on the inputs are each lower
than the index(es) of the output(s). Any intermediate
variables created by fanout are indexed to comply with
this convention. The primary outputs (one or more)
receive the highest indices for their variables. Proper
labeling has been shown in Figure 12.3.

12.2.1.3 Equation Reformulation. After constructing
all of the equations for a circuit, these equations are
solved to produce the Existence Function of the circuit.
The initial form of the equations is:

$$F_i (x_0, \ldots, x_{n-1}, x_n, \ldots, x_p) =$$
$$G_i (x_0, \ldots, x_{n-1}, x_n, \ldots, x_p)$$

where: $i = 1, \ldots, k$ (for k equations)

$\{ x_i \mid i = 0, 1, \ldots, n-1 \}$ the set of n
known variables (in this paper,
these are all of the circuit
primary inputs)

$\{ x_i \mid i = n, \ldots, p \}$ the set of m unknown
variables (in this paper, these
are all of the circuit primary
outputs, internal variables, and
fanout of primary inputs)

$p = n - 1 + m$ (when the starting index is
zero)

The formalism is usually discarded and the form noted as
simply:

$$F = G$$

These equations express the validity requirements which must be satisfied for the function to be true, that is, for the circuit output to be logical '1'. Therefore:

$$(F = G) <==> (y = 1)$$

where y denotes the output.

Equations in the form (F = G) may be rewritten as:

$$\overline{F}G + F\overline{G} = 0$$

which expresses the validity requirements for the complement function. Therefore:

$$(\overline{F}G + F\overline{G} = 0) <==> (y = 0)$$

The terms of the rewritten equations are either in the form $\overline{F}G$ or $F\overline{G}$ and are referred to as the Terms of Y, where $Y = \overline{y}$.

12.2.1.4 Generating the Existence Function. A system of equations in the form $(\overline{F}G + F\overline{G} = 0)$ are solved by the Boolean Analyzer when it is operated in the binary mode. The processing will result in the cancelling of all points in the logical space of the system of equations which are covered by at least one of the Terms of Y. This actually "cancels" the points of the Existence Function of Y, leaving its complement function, the Existence Function of y, in the memory.

For convenience, the Existence Function can be given the form of the Discriminant of the system of equations. This form uses all known variables as the independent (horizontal) index, and all unknown variables as the dependent (vertical) index.

The Existence Function contains information about all behavioral properties of the circuit. Fault testing problems are solvable by either: (1) Adding equations to the system of equations of the circuit prior to generating the Existence Function; or (2) performing further processing (with software) on the Existence Function after it is generated.

To demonstrate the generation of an Existence Function, Figure 12.4 details the procedure for Circuit 1 of Figure 12.3. The canceled points or squares represent the Existence Function of Y. The remaining points are the desired points of the Existence Function of y. The Function is repeated at the top of Figure 12.5.

12.2.2   Deriving the Test Sequence

12.2.2.1   Formation Rules. A Test Sequence is actually a sequence of tests or input vectors which are represented by the indices of the known variables of selected points. These points are the "one" points of the Existence Function which are interconnected by "links" or "bars".

A "link" or "bar" is formed by connecting any two points in the logical space of the Existence Function of a system which satisfy both of the following conditions: (1) The two points are logical distance one apart in the space of the known (horizontal) axis (note: Logical distance one is defined as the distance between two points in a logical space whose expression as Minterms differ in one and only one variable); (2) the two points are such that they differ in at least one observable output variable. The observable outputs are the primary outputs.

Figure 12.5 presented the Existence Function of Circuit 1 of Figure 12.3. The lower map shown in Figure 12.5b is the Existence Function redrawn with some of the possible links added, specifically, those links formed when the input variable x_1 or x_2 is used as the horizontal measure of logical distance.

12.2.2.2   Chain Selection. When all of the possible "bars" or "links" have been formed, there will be one or more "chains", which are sets of interconnected "links".

Repeating the equations of Circuit I, Figure 12.3:

$$x_5 = x_1 x_2 \qquad x_6 = x_3 x_4 \qquad x_7 = x_5 + x_6$$

These equations represent the relationships between the gate input and output lines. Each equation is in the form:

$$F = G$$

The equations are then rewritten using the relationship:

$$(\ F = G \ ) \ \text{<==>} \ (\ \bar{F}G + F\bar{G} = 0 \ )$$

into the following terms:

$\bar{F}G$: $\quad \bar{x}_5 x_1 x_2 \qquad \bar{x}_6 x_3 x_4 \qquad \bar{x}_7 x_5 \qquad \bar{x}_7 x_6$

$F\bar{G}$: $\quad x_5 \bar{x}_1 \qquad x_6 \bar{x}_3 \qquad x_5 \bar{x}_2 \qquad x_6 \bar{x}_4 \qquad x_7 \bar{x}_6 \bar{x}_5$

The Existence Function is formed by canceling all points in the logical space which are covered by at least one of these terms. For convenience, the circuit primary inputs are used as the known (horizontal) axis, the remaining variables, intermediate and outputs, are used as the unknown (vertical) axis.

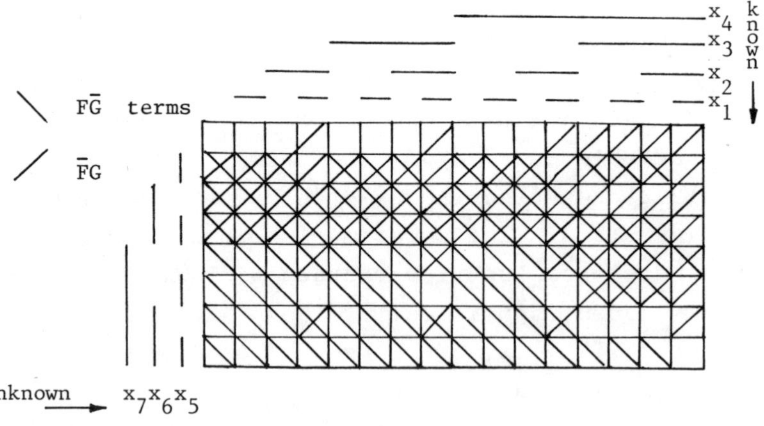

Figure 12.4.   Existence Function generation.

Following the cancelation of all points covered by one or more
terms, the remaining points in the logical space are the 'ONES'
of the Existence Function for the circuit. For this case
(Circuit I):

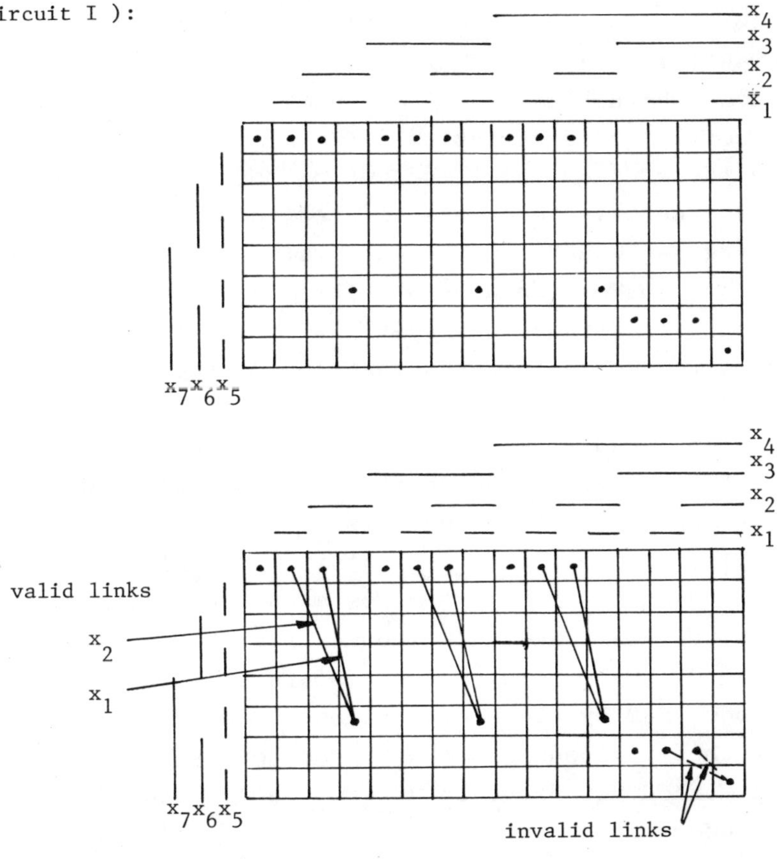

Partial interconnection of 'links'. Those shown are formed
from observing a change in the output variable x_7 when either
input variable x_1 or x_2 is changed, but not both.

Figure 12.5. Link formation.

The longest "chain" will produce the desired Test Sequence. There are cases where there is no longest chain. These cases are: (1) When a circuit is redundant, there are two or more sequences of equal length, and only one is necessary for fault detection; and (2) certain functions in which the terms of the function share no common true or complemented variables.

For those cases requiring two or more disconnected chains, a connecting sequence is used to join them. A "connecting sequence" must be the minimal length required to join the two Test Sequences while keeping its own points logical distance one apart. This latter requirement is to reduce the hazards that could otherwise be introduced by testing.

The completed linking for the example circuit is given in Figure 12.6. The short chains are discarded and the resulting Test Sequence is 5-7-6-14-10-11-9-13-5.

At first it was believed that it was sufficient to pick out a closed subset of points along the chain such that each type of link (each input variable variation) was included. By examining the results of other test set methods, and by close evaluation of the results produced by these methods and those produced by the Test Sequence method, it has been determined that the entire chain is necessary. The use of the entire chain of points ensures complete testing of both single and multiple faults. This includes multiple faults which do not have single-fault equivalences. The nature of the Sequence is such that each variable is tested for its ability to change value from '1' to '0' and from '0' to '1'.

12.2.2.3 Advantages of the Test Sequence. There are several desirable advantages offered by the Test Sequence.

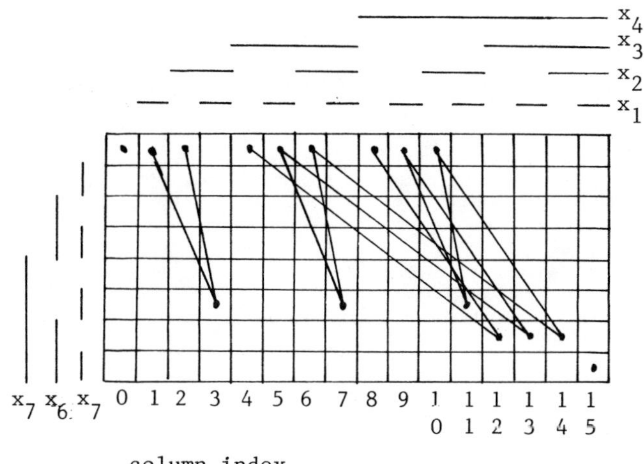

Existence Function of Circuit I, Figure 12.3, with all valid
links added. The links connecting points in columns 1-3-2
and those connecting 4-12-8 are not chained to a sufficient
length. The remaining chain connects 5-7-6-14-10-11-9-13-5.

Examination of the sequence of index values shows that when
this sequence of input values is applied to Circuit I, each
of the input variables will be tested independantly for a
change in value from 0 to 1 and again for a change in value
from 1 to 0. Each of the intermediate variables is also tested
in as thorough a manner and it is obvious from the rules of
link formation that the output variable (or variables)
is tested.

Figure 12.6. The Test Sequence.

255

The Test Sequence may be produced by the Boolean Analyzer, a hardware unit with a speed advantage over software approaches to test set generation.

An automated tester using the Test Sequence would not require continual resetting between tests and therefore throughput with this approach will be higher.

The requirement that there is no more than logical distance one between tests reduces the possibility of hazard introduction due to multiple test lead level changes.

The Test Sequence formation rules require that, upon the application of any test after the first, at least one observable output will change its logic level. This makes the detection of a failure logically straightforward.

The Test Sequence exercises every circuit variable through two-way logic level changes; that is, from '1' to '0' and from '0' to '1'. This ensures the detection of "temporarily correct" variables (those variables which change value and then become stuck at that value). As an aid, a diagram referred to as a Diagnostic Continuity diagram has been developed which represents graphically the variables which should change levels when the input changes from test to test (See Figure 12.7).

The Test Sequence is complete in its coverage of all detectable single faults. For those examples studied, the Test Sequence also covered the multiple faults which were not equivalent to a single fault.

The Test Sequence is closed, that is, it returns to the initial test. This reduces the resetting needed between circuits and may also have an advantage when intermittent fault detection is attempted.

There is also the possibility that the natural order of the Sequence will be advantageous in the detection of

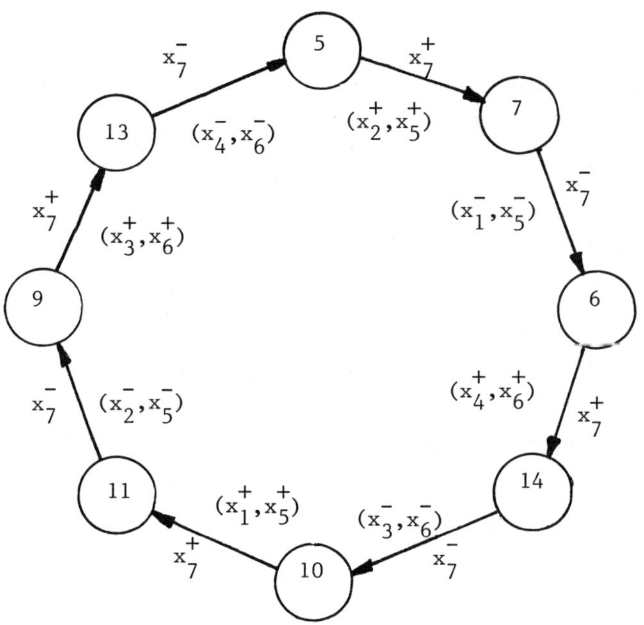

The Diagnostic Continuity diagram for Circuit I. The observable output is x_7 and is written separatly from the input and intermediate variables, shown in parenthesis. The nodes are labeled with the input vector.

Figure 12.7. Diagnostic Continuity diagram.

bridging faults. However, this topic has not been researched with the Sequence.

From the above advantages, it is concluded that coupling an automated tester of parallel input classification to the Test Sequence procedure will provide an economical and feasible solution to the fault detection problem for combinational circuits.

12.2.2.4 Possible Extension to Sequential Circuits. As an aside to the main line of research, a brief examination of sequential circuits has been made.

For sequential circuits, primary input variables are the known variables; primary output variables, intermediate variables, and feedback variables are the unknown variables. Formation of the linking for these cases requires a different approach, and this is an area of future research. It is believed that the Test Sequence approach can be extended to sequential circuits.

12.3 EXAMPLES

This section presents a few of the combinational circuits which were used in the research. The examples are ordered on the basis of their size.

12.3.1 Elementary Gates

The first example is a set of elementary gates, shown with their Existence Functions and Test Sequences in Figure 12.8. In accordance with their definitions by truth table, the OR and NOR gates are seen to have complementary Existence Functions, and they may be tested using the same Test Sequence. The same can be said of the AND and NAND gates. The Test Sequences in both cases contain the identical tests as specified by Eldred, and by others, for testing these individual gates, as was expected.

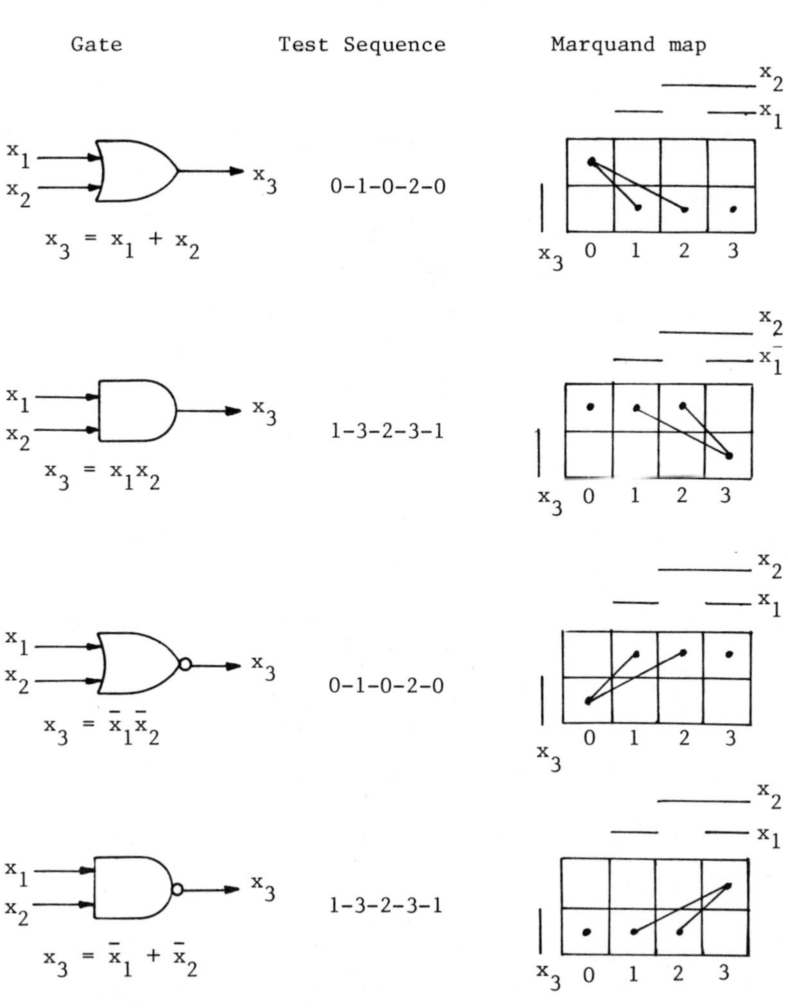

Figure 12.8. Test sequences for elementary gates.

12.3.2 Test Sequence vs. Boolean Difference

An example from the paper by Marinos is given in
Figure 12.9. This example was analyzed in detail by
Marinos using the Partial Boolean Difference approach,
and the minimized test set that he derived is given
on the second page of the figure. Both the system of
equations for the circuit, which has three fanout lines,
and the Terms of Y, where $Y = \overline{x}_{10} = \overline{f}$, are presented.

Due to the size of the full Existence Function
$(2^{11} = 2{,}048$ points), a scheme for computing the Test
Sequence from an ordered listing of its '1' points has
been used. The points are tabulated on the second page
with an asterisk (*) to denote the points chosen for
the Test Sequence (for this case, there is only one
chain).

The test set formed by the tests in the Sequence
and that found by Marinos are seen to be identical.

12.3.3 A Diagnostic Table

A fanout example is shown in Figure 12.10 with its
equations and, again, due to the size of the Existence
Function $(2^{12} = 4{,}096$ points), only the '1' points are
shown. The links have again been added to the value
table.

As a means of evaluating the Test Sequence, a Diagnostic Table, similar to a fault table, was designed and
is presented in Figure 12.11. The links are grouped as
row labels according to the input variable which alters
value, and the indices of all circuit variables are used
as the column headings. A '+' indicates that a variable
changes its logical value from '0' to '1'; a '-' means
that the reverse occurs; and a blank means that no change
has occurred (the links are shown for one direction only).
This is a tabular version of the Diagnostic Continuity
diagram.

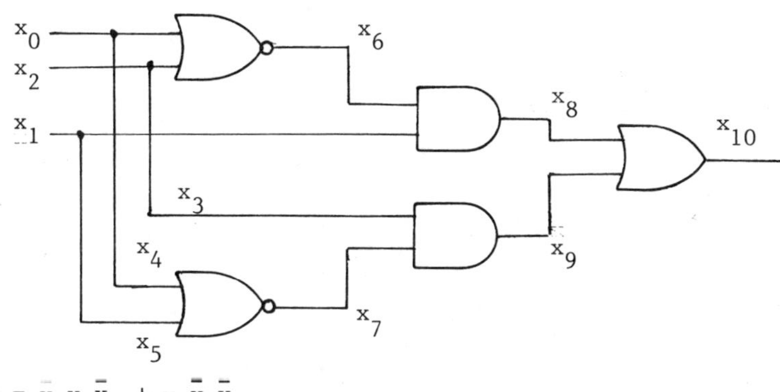

$$x_{10} = f = \bar{x}_2 x_1 \bar{x}_0 + x_2 \bar{x}_1 \bar{x}_0$$

System equations:

$$x_6 = \bar{x}_0 \bar{x}_2 \qquad x_7 = \bar{x}_4 \bar{x}_5 \qquad x_8 = x_6 x_1 \qquad x_3 = x_2$$

$$x_9 = x_3 x_7 \qquad x_{10} = x_8 + x_9 \quad x_4 = x_0 \qquad x_5 = x_1$$

The Terms of Y:

$$\bar{x}_6 \bar{x}_0 \bar{x}_2 \qquad \bar{x}_6 x_0 \qquad \bar{x}_6 x_2 \qquad \bar{x}_7 \bar{x}_4 \bar{x}_5 \qquad \bar{x}_7 x_4 \qquad \bar{x}_7 x_5$$

$$\bar{x}_8 x_6 x_1 \qquad \bar{x}_8 \bar{x}_6 \qquad \bar{x}_8 \bar{x}_1 \qquad \bar{x}_9 x_3 x_7 \qquad \bar{x}_9 \bar{x}_3 \qquad \bar{x}_9 \bar{x}_7$$

$$\bar{x}_{10} x_8 \qquad \bar{x}_{10} x_9 \qquad \bar{x}_3 x_2 \qquad x_{10} \bar{x}_8 \bar{x}_9 \qquad x_3 \bar{x}_2 \qquad \bar{x}_4 x_0$$

The Existence Function is so large that it is not shown here.
Instead a table of one points is given and the Test sequence
derived from this in the same manner it would have be derived
from the Existence Function.

Figure 12.9. An example from Marinos.

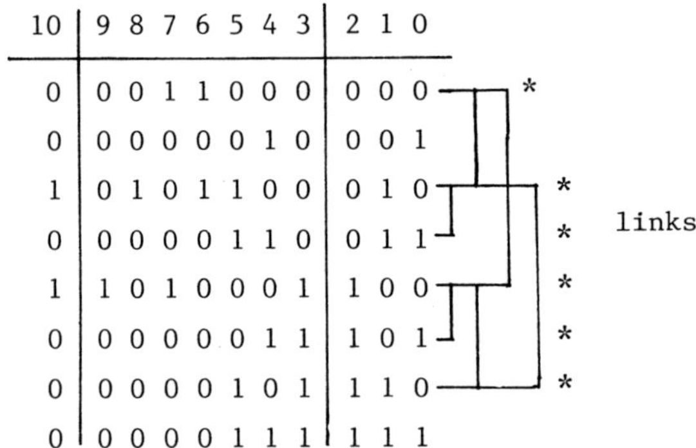

```
10 | 9 8 7 6 5 4 3 | 2 1 0
---+---------------+------
 0 | 0 0 1 1 0 0 0 | 0 0 0 ──┐  *
 0 | 0 0 0 0 0 1 0 | 0 0 1
 1 | 0 1 0 1 1 0 0 | 0 1 0 ─┐   *
 0 | 0 0 0 0 1 1 0 | 0 1 1 ─┘   *   links
 1 | 1 0 1 0 0 0 1 | 1 0 0 ─┐   *
 0 | 0 0 0 0 0 1 1 | 1 0 1 ─┘   *
 0 | 0 0 0 0 1 0 1 | 1 1 0 ──┘  *
 0 | 0 0 0 0 1 1 1 | 1 1 1
```

The Test Set from Marinos:

    0   4   2   6   5   3

The Test Sequence:

    3 − 2 − 0 − 4 − 5 − 4 − 6 − 2 − 3

Figure 12.9. (Con't)

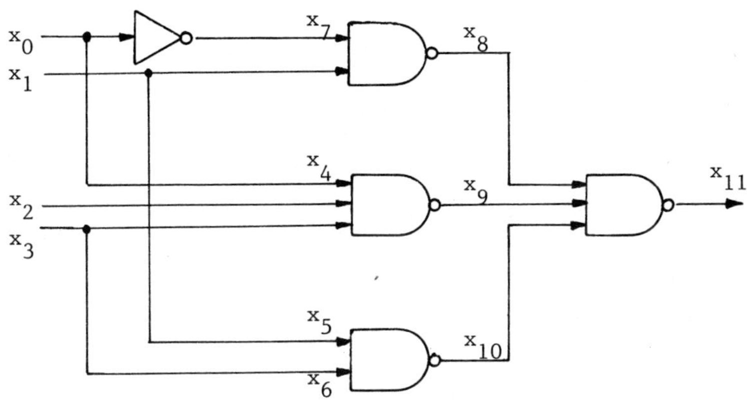

System equations:

$$x_7 = \bar{x}_0 \qquad x_4 = x_0 \qquad x_5 = x_1 \qquad x_6 = x_3$$

$$x_8 = \bar{x}_7 + \bar{x}_1 \qquad x_9 = \bar{x}_4 + \bar{x}_2 + \bar{x}_3 \qquad x_{10} = \bar{x}_5 + \bar{x}_6$$

$$x_{11} = \bar{x}_8 + \bar{x}_9 + \bar{x}_{10}$$

11	10 9 8 7 6 5 4	3 2 1 0
0	1 1 1 1 0 0 0	0 0 0 0
0	1 1 1 0 0 0 1	0 0 0 1
1	1 1 0 1 0 1 0	0 0 1 0
0	1 1 1 0 0 1 1	0 0 1 1
0	1 1 1 1 0 0 0	0 1 0 0
0	1 1 1 0 0 0 1	0 1 0 1
1	1 1 0 1 0 1 0	0 1 1 0
0	1 1 1 0 0 1 1	0 1 1 1
0	1 1 1 1 1 0 0	1 0 0 0
0	1 1 1 0 1 0 1	1 0 0 1
1	0 1 0 1 1 1 0	1 0 1 0
1	0 1 1 0 1 1 1	1 0 1 1
0	1 1 1 1 1 0 0	1 1 0 0
1	1 0 1 0 1 0 1	1 1 0 1
1	0 1 0 1 1 1 0	1 1 1 0
1	0 0 1 0 1 1 1	1 1 1 1

links

Figure 12.10.   An example for Bearnson and Carroll.

links	11	10	9	8	7	6	5	4	3	2	1	0	#	*
x0 links:														
2 - 3	-			+	-			+				+	#	*
6 - 7	-			+	-			+				+		*
12 - 13	+		-		-			+				+	#	*
x1 links:														
0 - 2	+		-				+				+		#	*
4 - 6	+		-				+				+			*
8 - 10	+	-	-				+				+			
9 - 11	+	-					+				+		#	*
12 - 14	+	-	-				+				+		#	*
x2 links:														
9 - 13	+		-				+						#	*
x3 links:														
3 - 11	+	-				+				+			#	*
5 - 13	+		-				+				+		#	*
7 - 15	+	-				+				+				*

(The rows 2-3 and 6-7 are bracketed together as "} or"; the rows 0-2 and 4-6 are likewise bracketed together as "} or".)

a. Diagnostic table.

The Test Set generated by Bearnson and Carroll:

$$\{\, {\textstyle\frac{2}{6}}, 11, 13, 12, 9, 7 \,\} \qquad \text{marked with } *$$

The Test Sequence by the link approach:

0-2-3-11-9-13-5-13-12-14-12-13-9-11-3-2-0

marked with #

Figure 12.11. Tabular sequence generation.

The Test Sequence points are shown by '#' and the
Test Set points generated by authors Bearnson and Carroll
are shown by '*'. One variable of the link pair appearing in the Test Set is sufficient for the '*' to appear
on that link. As can be seen, the methods produced
equivalent tests. The variation is between tests 7 and
3 which, when all variables are scanned for these two
input configurations, are seen to produce identical internal and output values.

12.3.4 Equivalent Circuits: Test Sequence vs.
        Kohavi's Maps

An example circuit is shown in Figure 12.12. and in
Figure 12.13.b. Figure 12.12 shows the labeling used in
Kohavi's paper and also TNC labeling. The equations are
given and the Existence Function '1' points are tabled
along with the decimal index of the input variables.
Both the Test Set derived by Kohavi and the Test Sequence
are given. Note the use by the Test Sequence of two of
the three tests given as equivalents in the Test Set, 0
and 4, and the addition of 2 ($\overline{DCB}\overline{A}$) as a test.

BOOLE, an APL program which emulates the Boolean
analyzer generation of an Existence Function, has been
used as a "check". Figure 12.14 presents the output generated by BOOLE when the input terms listed are used.
Note that APL uses $\underline{E}$ for $\overline{E}$ and uses letters of the alphabet for the line labels rather than x_i (TNC labeling was
not used in the program due to memory size limitations).
The Test Sequence produced by this version is the same
as before. (Figure 12.12.)

The circuit is repeated with two others in Figure
12.13. Circuits a and b are equivalent when their Marquand
maps are compared; circuit c is a complementary circuit.

The BOOLE output for the analysis of Circuit c is
shown in Figure 12.15. Note that all three circuits

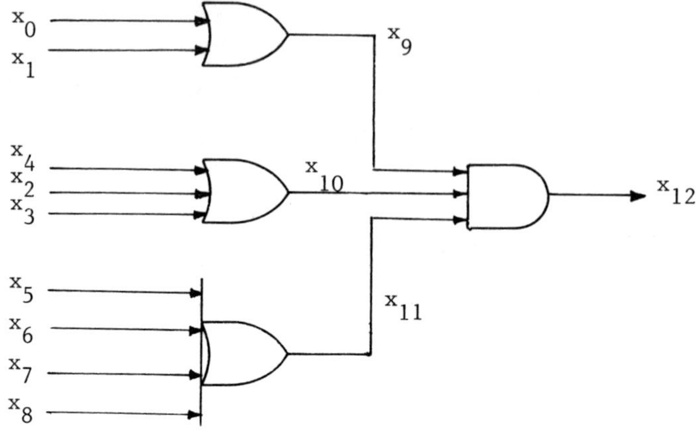

Equations:

$$x_4 = x_1 \quad x_8 = \bar{x}_3 \quad x_6 = \bar{x}_1 \quad x_5 = \bar{x}_0 \quad x_7 = \bar{x}_2$$

$$x_9 = x_0 + x_1 \quad x_{10} = x_2 + \bar{x}_3 + x_4 \quad x_{11} = x_5 + x_6 + x_7 + x_8$$

$$x_{12} = x_9 x_{10} x_{11}$$

12	11	10	9	8	7	6	5	4	$\bar{3}$	2	1	0	index
0	1	0	0	0	1	1	1	0	0	0	0	0	8
0	1	0	1	0	1	1	0	0	0	0	0	1	9
1	1	1	1	0	1	0	1	1	0	0	1	0	10
1	1	1	1	0	1	0	0	1	0	0	1	1	11
0	1	1	0	0	0	1	1	0	0	1	0	0	12
1	1	1	1	0	0	1	0	0	0	1	0	1	13
1	1	1	1	0	0	0	1	1	0	1	1	0	14
0	0	1	1	0	0	0	0	1	0	1	1	1	15
0	1	1	0	1	1	1	1	0	1	0	0	0	0
1	1	1	1	1	1	1	0	0	1	0	0	1	1
1	1	1	1	1	1	0	1	1	1	0	1	0	2
1	1	1	1	1	1	0	0	1	1	0	1	1	3
0	1	1	0	1	0	1	1	0	1	1	0	0	4
1	1	1	1	1	0	1	0	0	1	1	0	1	5
1	1	1	1	1	0	0	1	1	1	1	1	0	6
1	1	1	1	1	0	0	0	1	1	1	1	1	7

Test sequence: 2-0-1-9-11-15-7-15-14-12-13-15-13-9-1-0-2

Test set: $\left\{ 14,\ 13,\ 11,\ 7,\ 1,\ 9,\ 15,\ \begin{matrix} 0 \\ 4 \\ 12 \end{matrix} \right\}$

Figure 12.12. An example from Kohavi and Kohavi.

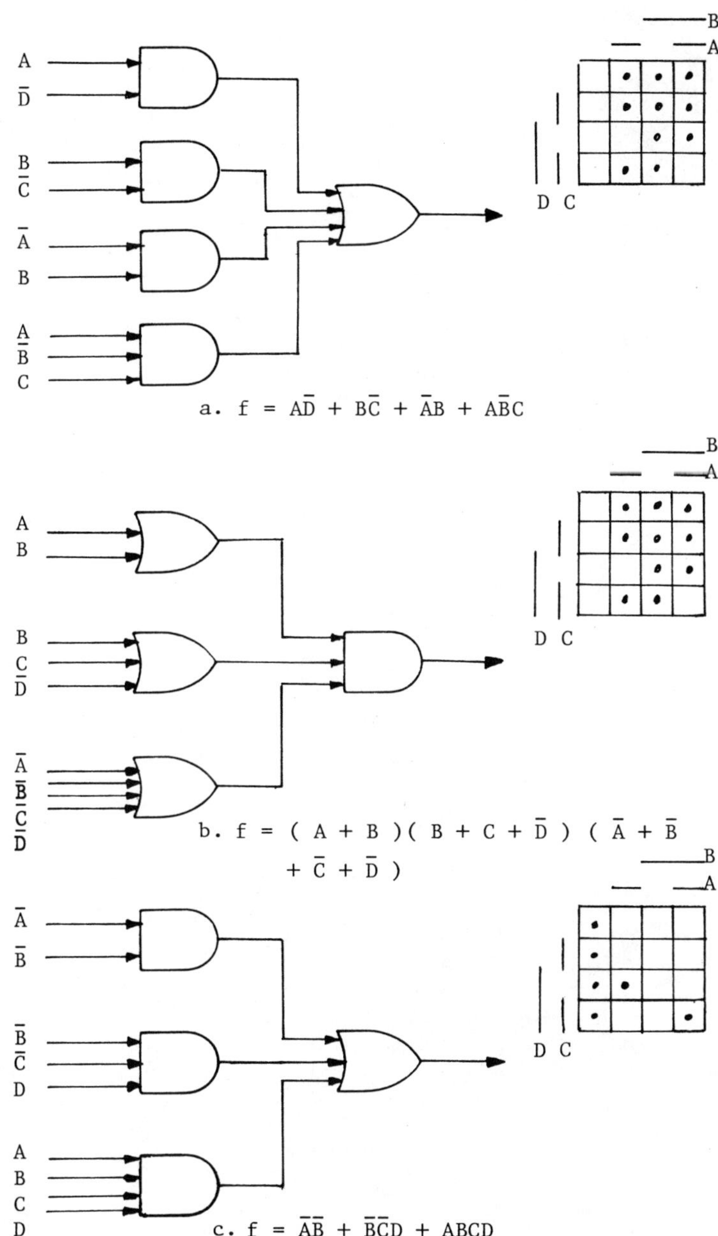

Figure 12.13.   Three example circuits from Kohavi and Kohavi.

Terms input to 'BOOLE':

$E\underline{A}$ $E\underline{B}$ $E\underline{AB}$ $F\underline{B}$ $F\underline{C}$ $F\underline{D}$ $FD\underline{BC}$ $G\underline{A}$ $G\underline{B}$ $G\underline{C}$ $G\underline{D}$ $GABCD$ $HEFG$
$H\underline{E}$ $H\underline{F}$ $H\underline{G}$

Discriminant output from BOOLE:

```
0 0 0 0 0 0 0 0 0 0 0 0 0 0 0 0
0 0 0 0 0 0 0 0 0 0 0 0 0 0 0 0
0 0 0 0 0 0 0 0 0 0 0 0 0 0 0 0
0 0 0 0 0 0 0 0 0 0 0 0 0 0 0 1
0 0 0 0 0 0 0 0 1 0 0 0 0 0 0 0
0 0 0 0 0 0 0 0 0 1 0 0 0 0 0 0
1 0 0 0 1 0 0 0 0 0 0 0 1 0 0 0
0 0 0 0 0 0 0 0 0 0 0 0 0 0 0 0
0 0 0 0 0 0 0 0 0 0 0 0 0 0 0 0
0 0 0 0 0 0 0 0 0 0 0 0 0 0 0 0
0 0 0 0 0 0 0 0 0 0 0 0 0 0 0 0
0 0 0 0 0 0 0 0 0 0 0 0 0 0 0 0
0 0 0 0 0 0 0 0 0 0 0 0 0 0 0 0
0 0 0 0 0 0 0 0 0 0 0 0 0 0 0 0
0 0 0 0 0 0 0 0 0 0 0 0 0 0 0 0
0 1 1 1 0 1 1 1 0 0 1 1 0 1 1 0
```

Figure 12.14.   BOOLE output for Figure 12.12.

```
0 1 1 1 0 1 1 1 0 0 1 1 0 1 1 0
0 0 0 0 0 0 0 0 0 0 0 0 0 0 0 0
0 0 0 0 0 0 0 0 0 0 0 0 0 0 0 0
0 0 0 0 0 0 0 0 0 0 0 0 0 0 0 0
0 0 0 0 0 0 0 0 0 0 0 0 0 0 0 0
0 0 0 0 0 0 0 0 0 0 0 0 0 0 0 0
0 0 0 0 0 0 0 0 0 0 0 0 0 0 0 0
0 0 0 0 0 0 0 0 0 0 0 0 0 0 0 0
0 0 0 0 0 0 0 0 0 0 0 0 0 0 0 0
1 0 0 0 1 0 0 0 0 0 0 0 1 0 0 0
0 0 0 0 0 0 0 0 0 1 0 0 0 0 0 0
0 0 0 0 0 0 0 0 1 0 0 0 0 0 0 0
0 0 0 0 0 0 0 0 0 0 0 0 0 0 0 1
0 0 0 0 0 0 0 0 0 0 0 0 0 0 0 0
0 0 0 0 0 0 0 0 0 0 0 0 0 0 0 0
0 0 0 0 0 0 0 0 0 0 0 0 0 0 0 0
```

The same test sequence is produced for each of the circuits shown in Figure 12.13.

Figure 12.15.   BOOLE output for Figure 12.13.c.

have the same Test Sequence. These and other examples
support the idea proposed by Akers and others that equivalent circuits, i.e. various implementations of a function, can be tested by the same Test Set. For the circuits shown, it is hypothesized that complementary equivalents, i.e. the various implementations of the complement of a function, can also be tested by this Test Set.

12.3.5 Multiple Faults

An example from the paper by Yau and Tang on multiple
faults appears in Figure 12.16. The Test Set they generated
for multiple fault detection for this circuit is seen to
contain the same points as appear in the Test Sequence.
The Diagnostic Continuity diagram for this circuit is
given in Figure 12.17.

The classic approach to Test Set minimization uses
a Fault Table and treats the minimization of the fault
test set as a coverage problem. (Note that the latest
algorithm for the Boolean analyzer is the Coverage algorithm.) The fault table for the example is given in
Figure 12.18. It should be noted that the test set thus
generated omitted 7 as a necessary test.

12.4 SUMMARY

Many other methods for test set generation exist
in the literature. Most of these become unmanageable
for large circuits.

The main objective of present research has been to
find efficient programmable algorithms. Of the existing
methods which were studied, Roth's D-algorithm program
set appears to be the most complete.

While the Test Sequence presented is programmable,
it is also possible to produce the results using the
Boolean Analyzer. The method is therefore a hardware
solution to the fault detection problem.

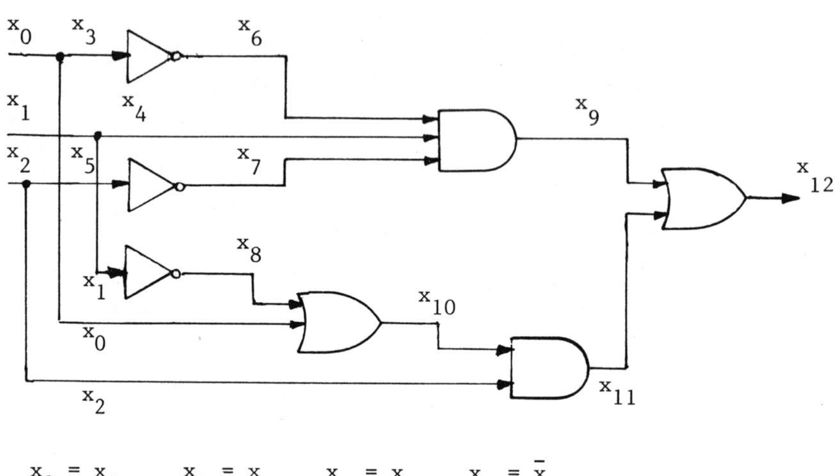

$$x_3 = x_0 \qquad x_4 = x_1 \qquad x_5 = x_2 \qquad x_6 = \bar{x}_3$$

$$x_7 = \bar{x}_5 \qquad x_8 = \bar{x}_1 \qquad x_9 = x_4 x_6 x_7$$

$$x_{10} = x_0 + x_8 \qquad\qquad x_{11} = x_2 x_{10} \qquad\qquad x_{12} = x_9 + x_{11}$$

12	11	10	9	8	7	6	5	4	3	2	1	0
0	0	1	0	1	1	1	0	0	0	0	0	0
0	0	1	0	1	1	0	0	0	1	0	0	1
1	0	0	1	0	1	1	0	1	0	0	1	0
0	0	1	0	0	1	0	0	1	1	0	1	1
1	1	1	0	1	0	1	1	0	0	1	0	0
1	1	1	0	1	0	0	1	0	1	1	0	1
0	0	0	0	0	0	1	1	1	0	1	1	0
1	1	1	0	0	0	0	1	1	1	1	1	1

links

Test sequence: 0–2–3–7–6–2–6–4–0

Test set for multiple fault detection: { 0, 4, 2, 6, 3, 7 }

Figure 12.16. Multiple fault testing example
from Yau and Tang.

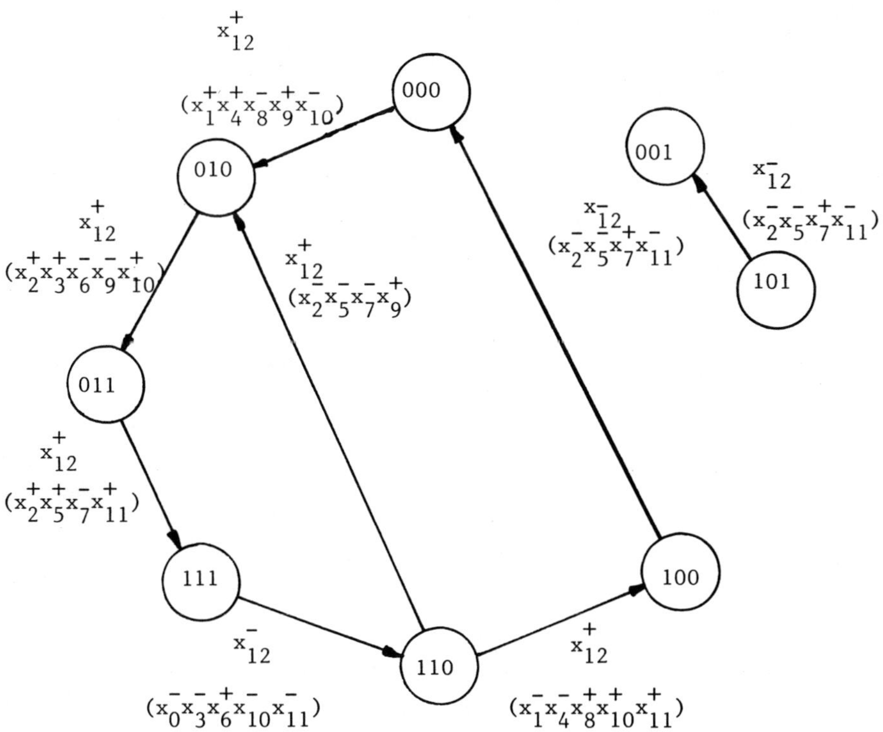

Test sequence: 0-2-3-7-6-2-6-4-0

The links are shown in only one direction for simplicity.

Figure 12.17. Diagnostic continuity diagram for Figure 12.16.

```
Single
Fault                   Input
Locations    SAX   0 1 2 3 4 5 6 7
       0      1    x x * x x x * x
              0    x x x * x x x *

       1      1    * x x x * x x x
              0    x x * x x x * x

       2      1    * * * * X X X X
              0    x x x x * * * *

       3      1    x x * x x x x x ← 2
              0    x x x * x x x x ← 3

       4      1    * x x x x x x x ← 0        |  *  test
              0    x x * x x x x x             |
                                               |  x  no test
       5      1    x x * x x x x x
              0    x x x x x x * x ← 6

       6      1    x x x * x x x x
              0    x x * x x x x x

       7      1    x x x x x x * x
              0    x x * x x x x x

       8      1    x x x x x x * x
              0    x x x x * x x x ← 4

       9      1    * * x * x x * x
              0    x x * x x x x x

      10      1    x x x x x x * x
              0    x x x x * * x *

      11      1    * * x * x x * x
              0    x x x x * * x *

      12      1    * * x * x x * x
              0    x x * x * * x *
```

Test set from Fault table above for Single Stuck-at faults:

$$\{ \ 0, \ 2, \ 3, \ 4, \ 6 \ \}$$

Figure 12.18. Fault table test set generation for Figure 12.16.

The Test Sequence generation method is intended for use with any multilevel, combinational circuit. It will also perform stable state test generation for sequential circuits; further research is needed to define a complete procedure for sequential circuits. There has been sufficient indication from the results of the examples which have been studied to hypothesize that the generation of the Test Sequence does not require knowledge of the internal functional logic of a circuit. The number of variables which may be handled is presently limited to the number of variables which the Analyzer is designed to process, twenty-two. For circuits which fit the size restriction, a complete, minimial test sequence is produced.

References

Svoboda, Antonin. "Class Notes in Engineering 125A," Computer Science Department, School of Engineering and Applied Science, University of California, Los Angeles, 1966 and 1972.

_______________. "Proposal for Research in Automated Design of Switching Circuits," UCLA-P-1483-N, May 1967.

_______________. "Ordering of Implicants," IEEE Trans., EC-16(2):100-105, Feb. 1967.

_______________. "Synthesis of Logical Systems of a Given Activity," IEEE Trans,, EC-12(12):904-910, Dec. 1963.

_______________. "Boolean Analyzer," Information Processing 68. North Holland, Amsterdam, 1969, pp. 824-830.

_______________. "Parallel Processing in Boolean Algebra," IEEE Trans., C-22(9):848-851, Sept. 1973.

_______________. "Boolean Analyzer Solutions of Fundamental Logic Design Problems," 1974 (unpublished).

_______________. "Logical Systems and Spaces," Paper Distributed at the Symposium on Computers, Prague, Czechoslovakia, 1964.

_______________. "Some Applications of Contact Grids," Proceedings, International Symposium on the Theory of Switching, Harvard U. Press, Cambridge, Mass., 1959, pp. 293-305.

_______________. "Logical Instruments for Teaching Logical Design," IEEE Trans. on Educ., E-12(4):262-273, Dec. 1969.

_______________. "The Concept of Term Exclusiveness and Its Effect on the Theory of Boolean Functions," JACM, 22(3):425-440, July 1975.

REFERENCES

DeVries, Ronald C., and Antonin Svoboda. "Multiple-Output Optimization with Mosaics of Boolean Functions," IEEE Trans., C-24(8):777-785, Aug. 1975.

Marquand, Allan. "On Logical Diagrams for n Terms," Philosophical Magazine, V. 12, pp. 266-270, 1881(year).

Karnough, M. "The Map Method for Synthesis of Combinational Logic Circuits," AIEE., pp. 593-599, Nov. 1953.

Bearnson, L. W. and C. C. Carroll. "On the Design of Minimal Length Fault Tests for Combinational Circuits," IEEE Trans., C-20(11):1353-1356, Nov. 1971.

Bouricius, W. G., et al. "Algorithms for Detection of Faults on Logic Circuits," Digest of Papers of the 1971 International Symposium on Fault Tolerant Computing, Pasadena, CA, March 1971 (IEEE pub # 71-C-6-C), pp. 5-8.

Kohavi, Zvi, and DeWayne Spires. "Designing Sets of Fault Detection Tests for Combinational Logic Circuits," IEEE Trans., C-20(12):1463-1469, Dec. 1971.

Kohavi, Igal and Zvi Kohavi. "Detection of Multiple Faults in Combinational Logic Networks," IEEE Trans., C-21(6):556-568, June 1972.

Marin, Miguel A. "Investigation of the Field of Problems for the Boolean Analyzer," Computer Science Department, University of California, Los Angeles, UCLA-ENG-6828, June 1968 (PhD Thesis).

White, D. E. "Fault Detection Through Parallel Processing in Boolean Algebra," UCLA-ENG-7504, March 1975 (PhD Thesis).

__________. "Test Sequence Alternative to Fault Detection," Seminar notes presented at Cal. State Polytechnic Institute, October 1974.

White, D. E., and Antonin Svoboda. "Fault Detection in Combinational Circuits: The Test Sequence," Proceedings, The Eigth Asilomar Conference on Circuits, Systems and Computers, Pacific Grove, CA., Dec. 1974. Pages UKN.

McCluskey, E. J. "Minimization of Boolean Functions," Bell Syst. Tech. J., 35:1417-1444, 1956.

REFERENCES

Marinos, Peter N. "Derivation of Minimal Complete Sets of Test Input Sequences Using Boolean Differences," IEEE Trans., C-20(1):25-32, Jan. 1971.

Sellers, F. F., and M. Y. Hsiao, and L. W. Bearnson. "Analyzing Errors with the Boolean Difference," IEEE Trans., C-17(4):676-683, July 1968, Correction, C-18(4):381, April 1969.

Yau, Stephen S., and Yu-Shan Tang. "An Efficient Algorithm for Generating Complete Test Sets for Combinational Logic Circuits," IEEE Trans., C-20(11):1245-1251, Nov. 1971.

Armstrong, D. B. "On Finding a Nearly Minimal Set of Fault Detection Tests for Combinational Logic Nets," IEEE Trans., EC-15(2):66-73, Feb. 1966.

Hsiao, M. Y., and Dennis K. Chia. "Boolean Difference for Fault Detection in Asynchrnous Sequential Machines," IEEE Trans., C-20(11):1356-1361, Nov. 1971.

Friedman, Arthur D., and P. R. Menon, Franklin F. Kho, ed. Fault Detection in Digital Circuits. Computer Applications in Electrical Engineering Series, Prentice-Hall, Englewood Cliffs, New Jersey, 1971.

Friedman, Arthur D. "Fault Detection in Redundant Circuits," IEEE Trans., C-16(2):99-100, Feb. 1967.

______________________. "Diagnosis of Short Circuit Faults in Combinational Circuits," IEEE Trans., C-23(7):746-752, July 1974.

Avizienis, Algirard. "Fault Tolerant Computers: Theory and Design of Ultra Reliable (Fault Tolerant) Computers: Protective Redundancy, Diagnosis, Self Repair Techniques," Class Notes, UCLA Seminar ENG 298/ENG 867.7 , March 1971.

Howell, T. H. Design Criteria for Error Detecting and Error Correcting Codes in Memory Subsystems, (February 1972 PhD Thesis) to be published by Garland.

Basham, Gary R. "New Error Correcting Technique for Solid State Memories Saves Hardware," Computer Design, pp. 110-113, Oct. 1976.

REFERENCES

Berger, Isreal, and Zvi Kohavi. "Fault Detection in Fan-out Free Combinational Networks," IEEE Trans., C-22(10):908-914, October 1973.

White, D. E., et al. "Introduction to Designing with the Am 2900 Family [bit-slice microprogrammed control devices]," Class Notes for AMD, Inc. Seminar ED2900A, June 1978.

Muroga, Saburo, and Hung Chi Lai. "Minimization of Logic Networks Under a Generalized Cost Function," IEEE Trans., C-25(9):893-907, Sept. 1976.

APL Program Index

KEY NAME   PAGE

CHAPTER 3... 34
     DECIDONT.. 35
     DECIMIN... 35
     BUILDIF... 36
     CHART... 36
     COMB.. 36
     DEGENERATION.. 36
     DISCRIMINANT.. 37
     EQUATION.. 37
     FFALSE.. 37
     FLIST... 38
     FORM.. 38
     FSTOR... 38
     FTRUE... 39
     LOGIC... 39
     MIN... 39
     MINIMA.. 40
     NEWORDER.. 41
     PRIMIMPLICANT... 42
     SOLVE... 42
     SPACE... 43
     TABLE... 44
     X... 44
     Y... 44
     Z... 44
     X... 44
     Y... 44
     Z... 44
     ΔFFALSE... 45
     ΔFTRUE.. 45
     ΔSYSTEM... 45

KEY NAME   PAGE

... TABLE... 45

CHAPTER 5.. 62
     AMPL.. 63
     BRANCH.. 64
     COMB.. 64
     COST.. 64
     CPU... 64
     COMMENT... 64
     DEFIN... 65
     DESIGN.. 65
     EXAMPLE... 66
     EXTEND.. 67
     FORWARD... 67
     DEL... 67
     FUN... 68
     GRAF.. 68
     IMPAS... 68
     INDX.. 69
     LIST.. 69
     MARC.. 69
     MINIMUM... 69
     MLUV.. 71
     OPTIMUM... 71
     ORDER... 72
     PRIMS... 73
     IND... 74
     SBOL.. 74
     SCHEMATIC... 74
     STOR.. 74
     TRIAL... 75

CHAPTER 7.. 108
     BULL.. 109
     CCMB.. 109
     COMMENTS.. 109
     CONST... 110
     DISCRIMINANT.. 110
     EXECUTE... 110
     FORMULA... 111
     REF... 112